AF822067

Praxis Core 5733, 5723, and 5713 Practice Tests Collection: Math, Writing, and Reading Study Guide and Mathematics Workbook Preparation for the Academic Skills for Educators Examinations

Praxis Core 5733, 5723, and 5713 Practice Tests Collection: Math, Writing, and Reading Study Guide and Mathematics Workbook (Preparation for the Academic Skills for Educators Examinations)

ISBN: 978-1-949282-76-4

For information on bulk discounts, please contact Exam SAM via our website.

Drawings in this publication are for guidance only. They are not drawn to scale.

NOTE: Praxis® and Praxis Core® are registered trademarks of Educational Testing Service (ETS). This publication is not associated with or endorsed by ETS.

Table of Contents

Praxis Core Math Test Format

The Praxis Core Mathematics Test 5733 contains the following types of questions:

- Number and Quantity – 20 questions
- Data Interpretation and Representation, Statistics, and Probability – 18 questions
- Algebra and Geometry – 18 questions

There are 56 questions in total on the Praxis Core math exam.

You will take the exam on a computer, unless you have applied for an exemption.

If you take the test on the computer, an on-screen calculator will be provided for you.

You will have 90 minutes to take the Praxis Core Math Test (5733).

Number and quantity questions cover the following skills:

- Positive and negative integers – including how to perform addition, subtraction, multiplication, and division, as well as understanding how to order numbers from highest to lowest or lowest to highest
- Mathematical equivalents and representing a value in more than one way – for example, converting a fraction to a decimal or percentage
- Whole numbers and their properties
- Ratios and proportion
- Decimals and percentages
- Dividing fractions by fractions
- Finding common factors and multiples
- US and metric systems of measurement

Data interpretation and representation, statistics, and probability questions cover:

- Reading and interpreting the visual display of quantitative information
- Understanding the relationship between data representation and data
- Making inferences form given visual displays of data
- Working with data and data representation to solve problems
- Analyzing scatterplots
- Interpreting distributions
- Drawing conclusions based on the data provided
- Calculating mean, median, mode, and range
- Calculating basic probability
- Using random sampling to draw conclusions about data
- Understanding probability models
- Understanding the difference between statistical and non-statistical questions

Algebra and geometry questions cover:

- Problem solving techniques for algebraic expressions, as well as real-life problems
- Equations and inequalities
- Solving expressions that contain one variable
- Finding equivalent expressions
- Narrative and mathematical representations of data
- Linear equations
- Simple quadratics
- Graphs
- Solving problems on geometric shapes

    - Triangles, rectangles, squares, and circles
    - Real-life problems on angles, area, perimeter, and box volume
    - Describing or drawing figures
    - Interpreting relationships between figures
- Pythagorean theorem
- Understanding congruency and similarity

Most of the questions will be multiple-choice, but there will be some numeric entry questions.

For the numeric entry questions, you have to make a calculation and provide the answer on your own. You will see a small box in which to place your answer, but you will not be provided with any answer choices.

You should also be aware that on some of the multiple-choice questions, you will be allowed to choose more than one response.

How to Use This Publication

Practice Test 1 of the math study guide is in workbook format, with the questions organized by skill area and with study tips and solutions after each question.

The workbook format of practice test 1 will help you learn the strategies and formulas that you need to answer all of the types of questions on the actual Praxis Core Math Test.

You may wish to time yourself as you do the remaining practice tests in this book, allowing yourself 90 minutes for each exam. This will help to simulate the conditions of the actual test.

Praxis Core Math Practice Set 1 with Workbook

Number and Quantity Problems:

1) Which of the following shows the numbers ordered from least to greatest?

A) $-{}^{1}/_{4}$, ${}^{1}/_{8}$, ${}^{1}/_{6}$, 1
B) $-{}^{1}/_{4}$, ${}^{1}/_{8}$, 1 , ${}^{1}/_{6}$
C) $-{}^{1}/_{4}$, ${}^{1}/_{6}$, ${}^{1}/_{8}$, 1
D) $-{}^{1}/_{4}$, 1 , ${}^{1}/_{8}$, ${}^{1}/_{6}$
E) 1 , ${}^{1}/_{6}$, ${}^{1}/_{8}$, $-{}^{1}/_{4}$

> In order to answer questions on ordering numbers from least to greatest or greatest to least, remember these principles: (a) Negative numbers are less than positive numbers; (b) When two fractions have the same numerator, the fraction with the smaller number in the denominator is the larger fraction.

The correct answer is A.

According to the principles above, $-{}^{1}/_{4}$ is less than ${}^{1}/_{8}$, ${}^{1}/_{8}$ is less than ${}^{1}/_{6}$, and ${}^{1}/_{6}$ is less than 1.

2) Consider the instructions in the chart below, and then answer the question that follows.

> Step 1: Begin with a number N.
> Step 2: Multiply the number by 3.
> Step 3: Subtract 4 from the result.
> Step 4: Divide the result by 8.

If the result of the calculation is 4, what is the value of N?

A) 1 B)2 C)12 D)64 E) 0.667

> Work backwards, using inverse operations. So, if you are asked to divide you multiply to get the solution. If you are asked to subtract, you add to get the solution, and so on.

The correct answer is C.

Step 4 says "divide by 8," so multiply our final result of 4 by 8 to perform the first inverse operation: $4 \times 8 = 32$

Step 3 says "subtract 4," so add 4 to our previous result to perform the inverse operation: $32 + 4 = 36$

Step 2 says "multiply by 3," so divide by 3 to perform the final inverse operation to solve: $36 \div 3 = 12$

3) A recent report states that 72.8% of the work for the shopping center is now completed, and it has taken 182 days to do so. If work continues at the same rate, what fraction of the project will be completed after 43 more days? Put the correct amounts in the spaces provided.

$$\frac{\boxed{}}{\boxed{}}$$

First you need to divide the number of days into the percentage in order to determine what percentage of the project is being completed each day.

The correct answer is $^{9}/_{10}$.

As mentioned above, divide the percentage by the number of days in order to determine what percentage of the project is being completed each day:

72.8% ÷ 182 days = 0.4 percent completed per day

For the numerator of the fraction, we need to add 43 more days to the current 182 days: 182 + 43 = 225

Then determine how many days are needed to complete the entire project for the denominator of the fraction.

Since a complete project would be 100% complete, we divide the percentage per day into 100% to get the total days needed:

100% ÷ 0.4 = 250 days

Finally, express this as a simplified fraction:

$$\frac{225}{250} \div \frac{25}{25} = \frac{9}{10}$$

4) The temperature on Saturday was 62° F at 5:00 PM and 38° F at 11:00 PM. If the temperature fell at a constant rate on Saturday, what was the temperature at 9:00 PM?

A) 58° F B) 54° F C) 50° F D) 46° F E)40° F

This question assesses your knowledge of performing operations on integers. Here, we have to perform the operations of subtraction, multiplication, and division.

The correct answer is D.

First of all, you need to determine the difference in temperature during the entire time period: 62 – 38 = 24 degrees less

Then calculate how much time has passed. From 5:00 PM to 11:00 PM, 6 hours have passed.

Next, divide the temperature difference by the amount of time that has passed to get the temperature change per hour.

24 degrees ÷ 6 hours = 4 degrees less per hour

To calculate the temperature at the stated time, you need to calculate the time difference.

From 5:00 PM to 9:00 PM, 4 hours have passed. So, the temperature difference during the stated time is 4 hours × 4 degrees per hour = 16 degrees less.

Finally, deduct this from the beginning temperature to get your final answer.

62° F – 16° F = 46° F

5) A baker multiplied a recipe by $^{1}/_{2}$ when he should have divided the recipe by 4. Which one of the operations on the erroneous result will correct the error?

A) Divide by $^{1}/_{2}$

B) Multiply by $^{1}/_{2}$

C) Multiply by 2

D) Divide by 4
E) Multiply by 8

For erroneous calculations that involve division, try to find an equivalent fraction for the division. The baker should have divided by 4. Dividing by 4 is equal to multiplying by what fraction?

The correct answer is B.

The baker should have divided the recipe by 4.

Dividing by four is the same as multiplying by $^1/_4$.

The baker erroneously multiplied by $^1/_2$, so he needs to multiply by $^1/_2$ again since $^1/_2 \times {}^1/_2 = {}^1/_4$.

6) A painter needs to paint 8 rooms, each of which have a surface area of 2000 square feet. If one bucket of paint covers 900 square feet, what is the fewest number of buckets of paint that must be used to complete all 8 rooms?
A) 3 B) 17 C) 18 D) 19 E) 20

This is a question that requires you to find the fewest multiples of an item. Be mindful of the words "fewest" and "greatest" in problems like this one, since it will normally be impossible to purchase a fractional part of the item in the question. Therefore, you will need to round your result up or down accordingly.

The correct answer is C.

For your first step, determine how many square feet there are in total: 2000 square feet per room × 8 rooms = 16,000 square feet in total

Then you need to divide by the coverage rate:

16,000 square feet to cover ÷ 900 square feet coverage per bucket = 17.77 buckets needed

It is not possible to purchase a partial bucket of paint, so 17.77 is rounded up to 18 buckets of paint.

7) Soon Li jogged 3.6 miles in $^3/_4$ of an hour. What was her average jogging speed in miles per hour?
A) 2.7 B) 4.0 C) 4.2 D) 4.6 E) 4.8

This problem involves the calculation of miles per hour. To solve the problem, divide the distance traveled by the time in order to get the speed in miles per hour.

The correct answer is E.

As stated previously, divide the distance traveled by the time in order to get the speed in miles per hour.

Remember that in order to divide by a fraction, you need to invert the fraction, and then multiply.

3.6 miles ÷ $^3/_4$ =

3.6 × $^4/_3$ =

(3.6 × 4) ÷ 3 =

14.4 ÷ 3 = 4.8 miles per hour

8) When 1523.48 is divided by 100, which digit of the resulting number is in the tenths place?
A) 1 B) 2 C) 3 D) 4 E) 5

This question assesses your understanding of place value. Remember that the number after the decimal is in the tenths place, the second number after the decimal is in the hundredths place, and the third number after the decimal is in the thousandths place.

The correct answer is B.

Perform the division, and then check the decimal places of the numbers. Divide as follows: 1523.48 ÷ 100 = 15.2348

Reading our result from left to right: 1 is in the tens place, 5 is in the ones place, 2 is in the tenths place, 3 is in the hundredths place, 4 is in the thousandths place, and 8 is in the ten-thousandths place.

9) The price of a certain book is reduced from $60 to $45 at the end of the semester. By what percent is the price of the book reduced?
A) 15% B) 20% C) 25% D) 33% E) 45%

This question asks you to perform a calculation in order to determine the percentage discount on an item. Divide the dollar value of the reduction by the original price to get the percentage.

The correct answer is C.

Determine the dollar amount of the discount.

$60 original price – $45 sale price = $15 discount

Then divide the discount by the original price to get the percentage of the discount.

$15 ÷ $60 = 0.25 = 25%

10) Mr. Rodriguez teaches a class of 25 students. Ten of the students in his class participate in drama club. In which graph below does the dark gray area represent the percentage of students who participate in drama club?

A)

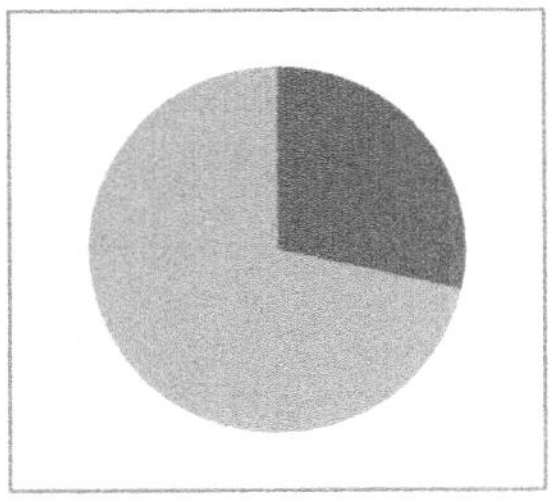

B)

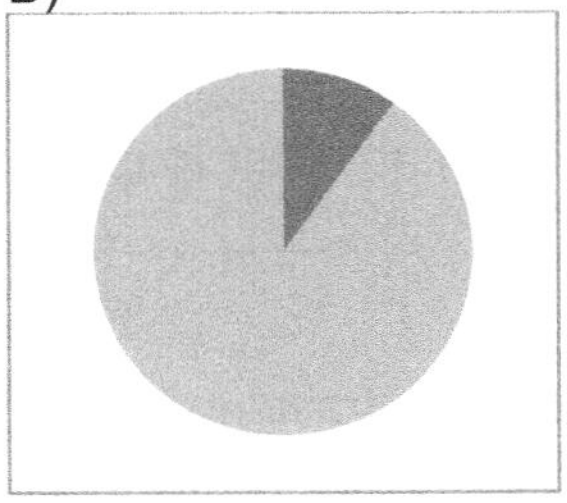

C)

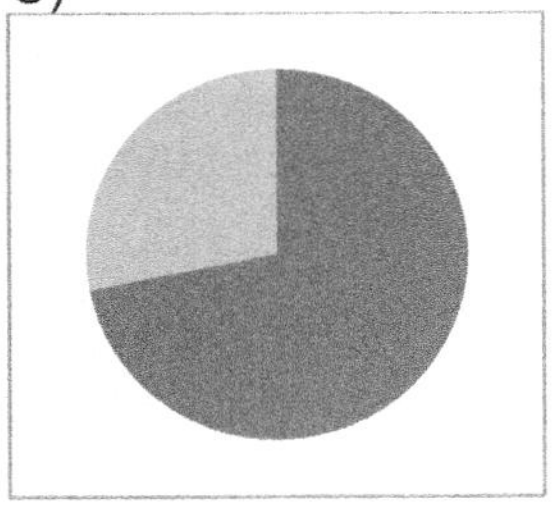

D)

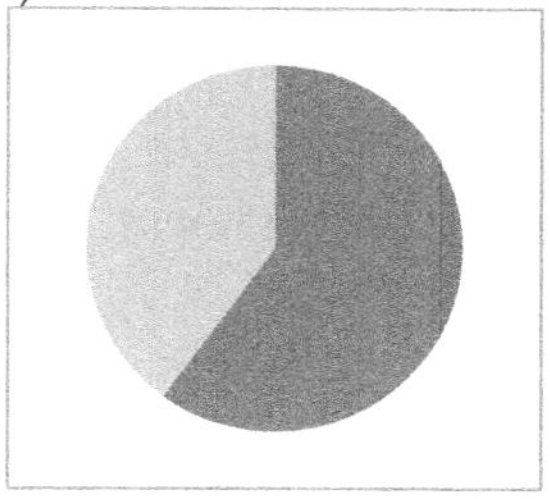

E)

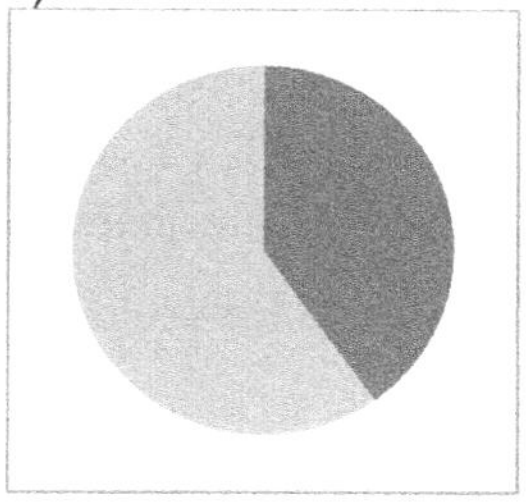

Questions like this one are asking you about how to express percentages graphically. Facts such as x students from y total students participate in a group can be represented as $^{x}/_{y}$.

The correct answer is E.

Ten out of 25 students participate in drama club.

First of all, express the relationship as a fraction: $^{10}/_{25}$

Then divide to find the percentage: $^{10}/_{25} = 10 \div 25 = 0.40 = 40\%$

Finally, choose the pie chart that has 40% of its area shaded in dark gray.

40% is slightly less than half, so you need to choose chart E.

11) The ratio of males to females in the senior year class of Carson Heights High School was 6 to 7. If the total number of students in the class is 117, how many males are in the class?

A) 48 B) 54 C) 56 D) 58 E) 63

Remember that a ratio can be expressed by using the word "to" or by separating the amounts in the subsets with a colon. So, our ratio is expressed as 6 to 7 or 6:7.

The correct answer is B.

For your first step, add the subsets of the ratio together: 6 + 7 = 13

Then divide this into the total: 117 ÷ 13 = 9

Finally, multiply the result from the previous step by the subset of males from the ratio: 6 × 9 = 54 males in the class

12) Results from a questionnaire administered to customers of a particular grocery store show that 4 out of 7 customers prefer toffee-flavored ice cream to coffee-flavored ice cream. Based on these results, if 1,540 customers purchased one of these two flavors of ice cream, how many of them would have purchased coffee-flavored ice cream?

A) 220 B) 420 C) 560 D) 660 E) 880

This is a question on a proportion, not on a ratio. Proportions often use the phrase "a out of b people." So, you need to do another step before you can get the ratio from the proportion.

The correct answer is D.

Be careful with the phrase “out of” in proportion questions like this one. We are given the phrase “4 out of 7.”

If 4 out of 7 prefer the toffee flavor, then the remaining 3 prefer the coffee flavor.

Accordingly, the ratio of flavor preference of toffee to coffee is 4 to 3.

So, after determining the ratio, your next step is to divide the total by 7: $1540 \div 7 = 220$

Then multiply this by 3 for the coffee-flavor preference: $220 \times 3 = 660$

13) A measurement of 116 feet is how many inches longer than a measurement of 36 yards?
A) 8 B) 80 C) 96 D) 960 E) 3,744

Convert yards to feet and then feet to inches. How many feet are in a yard? How many inches are in a foot? Refer to the appendix at the back of the book, if needed.

The correct answer is C.

First of all, convert 36 yards to feet: 36 yards $\times$ 3 feet in a yard = 108 feet

Then subtract this from 116 feet: $116 - 108 = 8$ feet

Then convert the feet to inches for your answer:

8 feet $\times$ 12 inches per foot = 96 inches

14) The graph below shows the relationship between the total number of hamburgers a restaurant sells and the total sales in dollars for the hamburgers. What is the sales price per hamburger?

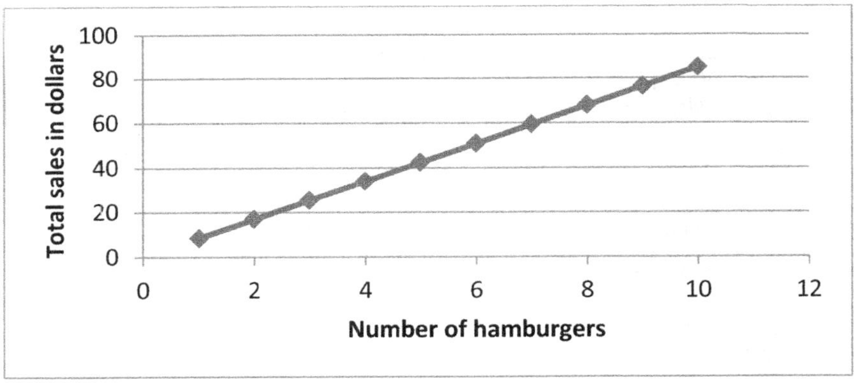

A) \$4.00 B) \$8.00 C) \$8.50 D) \$9.50 E) \$10.00

Although the question involves a graph, it is essentially a number and quantity question involving division. To solve the problem, look at the graph and then divide the total sales in dollars for a particular number of units by the quantity sold in order to get the price per unit.

The correct answer is C.

For ten hamburgers, the total price is \$85, so each hamburger sells for \$8.50: \$85 total sales in dollars $\div$ 10 hamburgers sold = \$8.50 each

15) In Brown County Elementary School, parents are advised to have their children vaccinated against five childhood diseases. According to the chart below, how many children were vaccinated against at least three diseases?

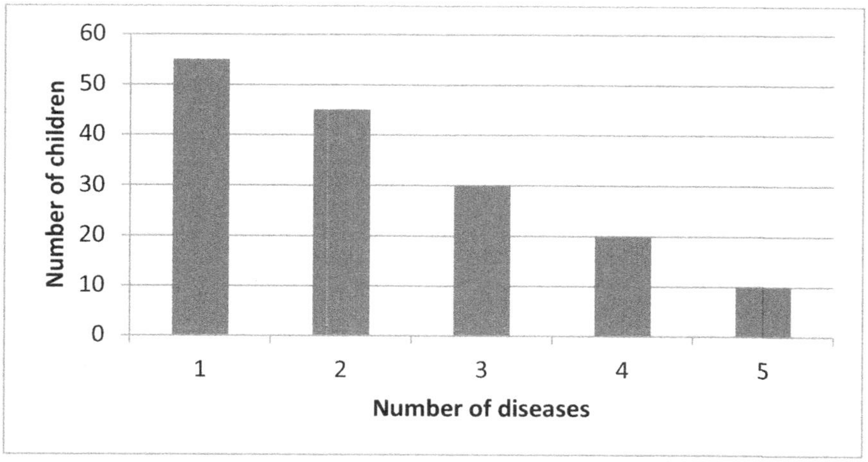

A) 30 B) 50 C) 60 D) 100 E) 130

For number and quantity questions that ask you to interpret bar graphs, you need to read the problem carefully to determine what is represented on the horizontal axis (bottom) and the vertical axis (left side) of the graph.

The correct answer is C.

The quantity of diseases is indicated on the bottom of the graph, while the number of children is indicated on the left side of the graph.

To determine the number of children that have been vaccinated against three or more diseases, we need to add the amounts represented by the bars for 3, 4, and 5 diseases: $30 + 20 + 10 = 60$ children

16) In a sequence of integers, the first number is N and each number in the sequence thereafter is half of the previous number. If the sixth integer in the sequence is 4, what is the first integer?

A) 256 B) 192 C) 128 D) 108 E) 64

When you are given the final term in a sequence and asked for a previous term, you need to divide by the size of the increment to solve. Here, our increment is one-half.

The correct answer is C.

Remember to divide by the size of the increment to solve. Again, our increment is one-half.

When we divide by one-half, we need to multiply by two since $1 \div \frac{1}{2} = 2$.

So, multiply like this to solve:

Term 6: 4; Term 5: $4 \times 2 = 8$; Term 4: $8 \times 2 = 16$; Term 3: $16 \times 2 = 32$; Term 2: $32 \times 2 = 64$; Term 1: $64 \times 2 = 128$

17) What is the greatest common divisor of the following integers? 204 and 272

A) 72 B) 68 C) 34 D) 24 E) 4

The greatest common divisor is the largest number that will divide into at least two other numbers.

The correct answer is B.

Here our greatest common divisor is 68 since $204 \div 68 = 3$ and $272 \div 68 = 4$.

If you feel stuck, divide by the first obvious number you can think of, and keep dividing until you can't do any further division.

For example, first you could divide by 4:

$204 \div 4 = 51$ and $272 \div 4 = 68$

Then see if you can divide further. Be sure to consider the prime numbers as possible divisors.

Both 51 and 68 from above are divisible by 17, so divide as shown:

$51 \div 17 = 3$ and $68 \div 17 = 4$

Then multiply the individual divisors together to get the greatest common divisor:

$4 \times 17 = 68$

18) Susan is trying to limit her caffeine intake. She has a coffee in the morning every third day and a glass of iced tea every fourth day in the afternoon. For how many of the next 90 days will Susan consume coffee and iced tea on the same day?

A) 7 B) 10 C) 12 D) 22 E) 30

For interval questions like this one, multiply the number of days of each interval together to get the day on which the intervals will coincide.

The correct answer is A.

Every twelfth day, Susan will have both drinks on the same day.

We get this result by multiplying 3 days for the coffee interval by 4 days for the tea interval: $3 \times 4 = 12$

Then we have to determine how many 12-day intervals there are in 90 days. We do this by dividing 90 by 12: $90 \div 12 = 7.5$

We don't count the half day, so the answer is 7.

19) Which of the equations provided below is sufficient to illustrate that the following statement is false?

If $x \times y$ can be divided by 6, then x is divisible by 6 or y is divisible by 6.

A) $4 \times 5 = 20$ B) $3 \times 4 = 12$ C) $6 \times 3 = 18$ D) $5 \times 6 = 30$ E) $8 \times 6 = 48$

This question is asking you to find counter-examples to a statement by using basic operations. First of all, rule out all answer choices that include a factor that is the same as the divisor stated in the facts of the question.

The correct answer is B.

Answers C, D, and E all multiply another number by 6. These factors of 6 are of course divisible by 6, which supports the statement above instead of disproving it.

Answer A is irrelevant since neither the factors nor the product is divisible by 6.

Answer B disproves the statement because 12 is divisible by 6, although neither 3 nor 4 is divisible by 6.

20) In a particular section of a library, the ratio of non-fiction books to fiction books is 7 to 9. The total number of fiction and non-fiction books in this section of the library is 128. If 10 fiction books and 14 non-fiction books are retired and removed from this section of the library, what fraction of the remaining books in this section are fiction? Put the correct amounts in the spaces provided.

[ ]

———

[ ]

> This is an advanced question on ratios. First, add the two parts of the ratio together. Then divide this result into the total number of books. Then multiply each part of the ratio by this result to get the number of each type of book.

The correct answer is 31 / 52.

Add the two parts of the ratio together: 7 + 9 = 16

Then divide this into the total number of books: 128 ÷ 16 = 8

Then multiply each part of the ratio by 8 to get the number of each type of book:

7 × 8 = 56 non-fiction books

9 × 8 = 72 fiction books

Then subtract the removals and find the new total:

56 – 14 = 42 non-fiction left

72 – 10 = 62 fiction left

42 + 62 = 104 total books left

Finally, express the fiction books as a fraction of the total: 62 / 104 = 31 / 52

Data Interpretation and Representation:

21) A mother has noticed that the more sugar her child eats, the more her child sleeps at night. Which of the following graphs best illustrates the relationship between the amount of sugar the child consumes and the child's amount of sleep?

A)

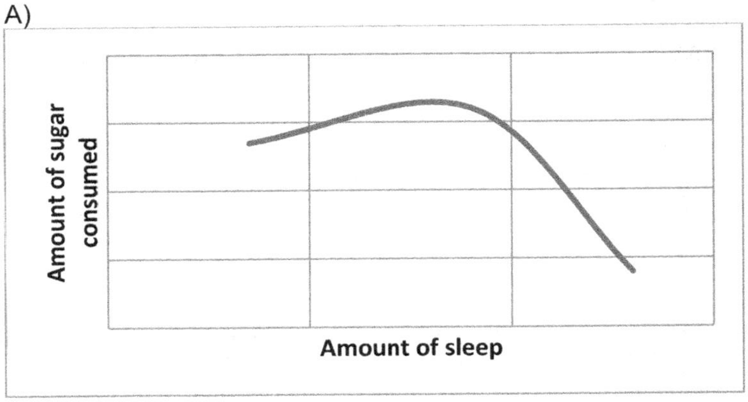

B)

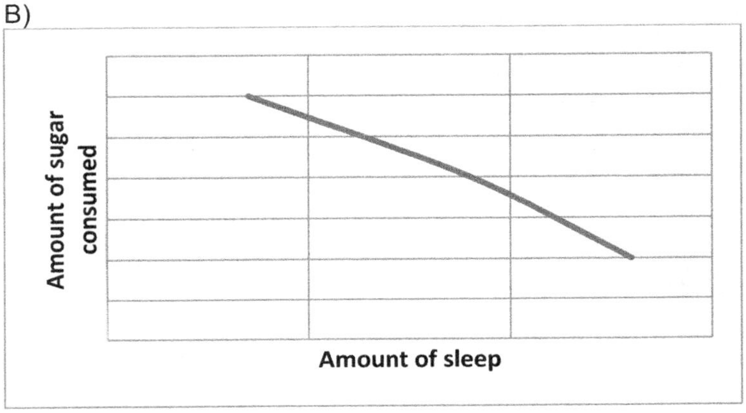

C)

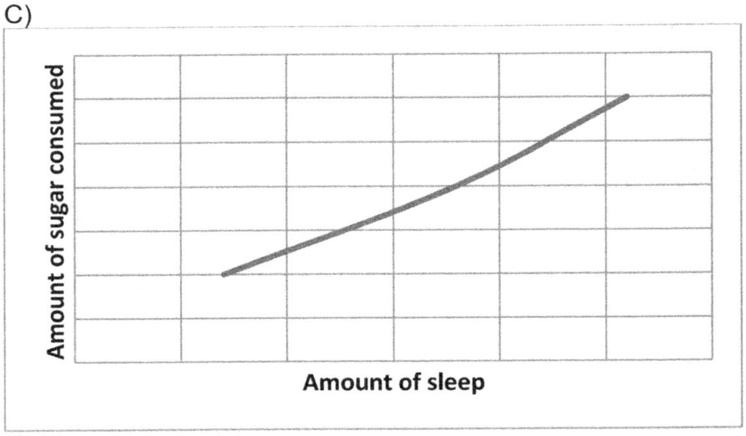

D)

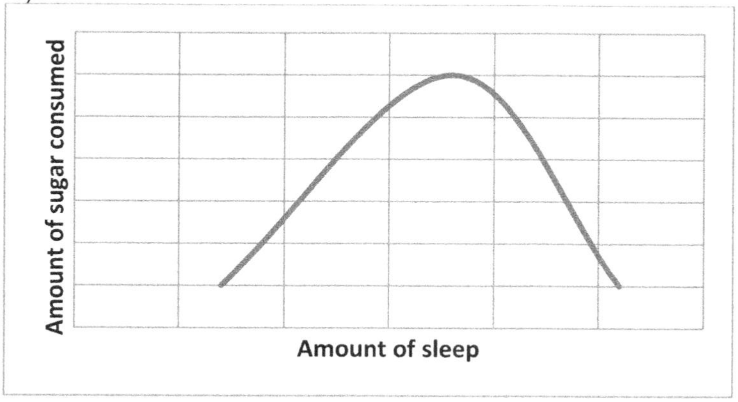

E)

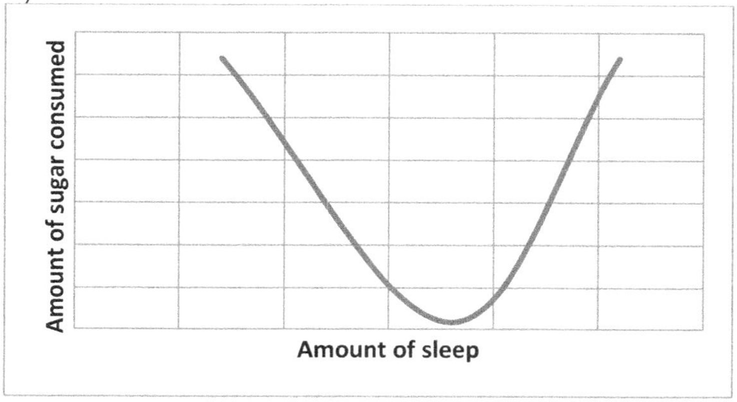

Your exam will have problems like this one that show graphs or scatterplots that represent the relationship between two variables. Be sure that you know the difference between positive linear relationships and negative linear relationships for the exam. In a positive linear relationship, an increase in one variable is correlated to an increase in the other variable, meaning that the line will point upwards from left to right.

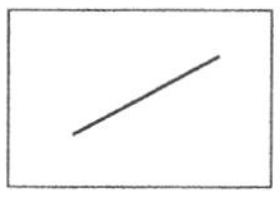

In a negative linear relationship, an increase in one variable is correlated to a decrease in the other variable, meaning that the line will point downwards from left to right.

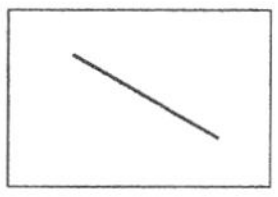

The correct answer is C.

As the quantity of sugar increases, the amount of sleep also increases.

A positive linear relationship therefore exists between the two variables. This is represented in chart C since the amount of sleep is greater when the amount of sugar consumed is higher.

22) Which one of the scatterplots below most strongly suggests a negative linear relationship between x and y?

A)

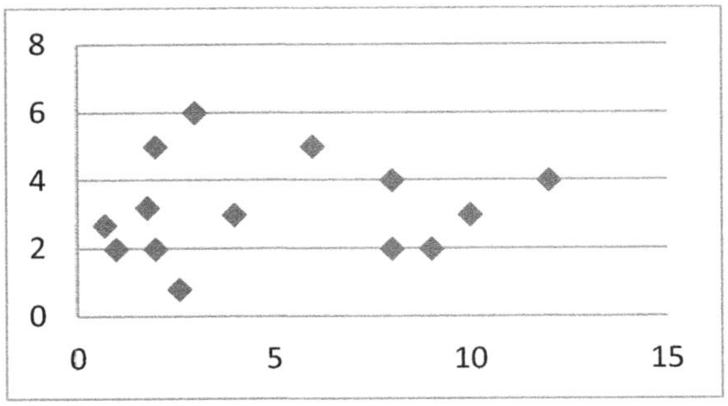

B)

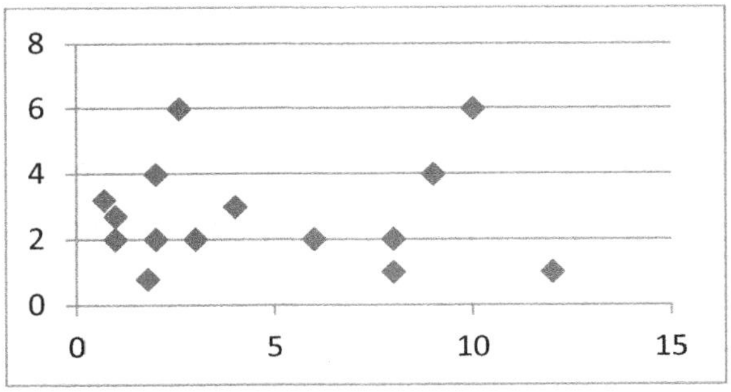

C)

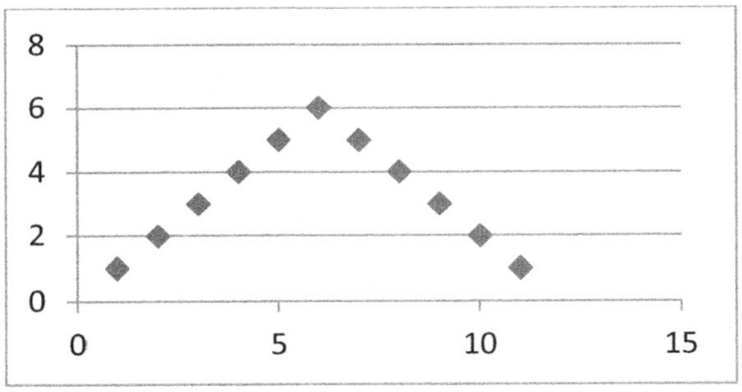

D)

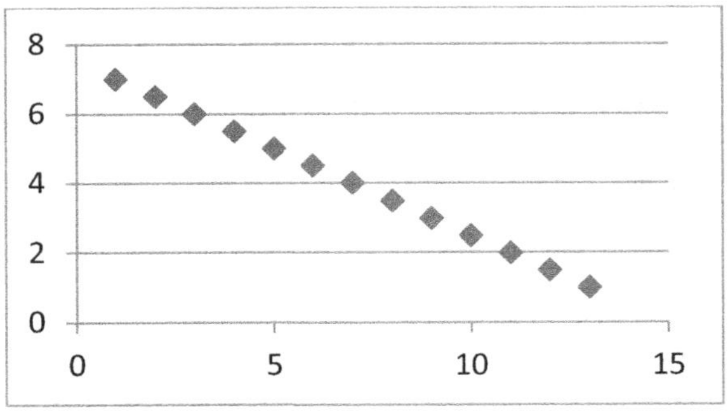

E)

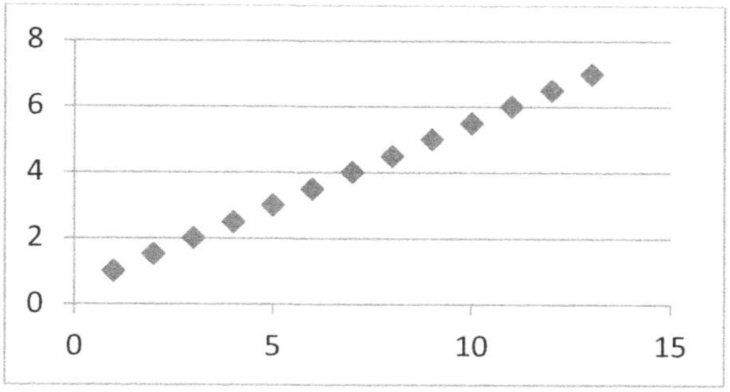

Your exam will have problems containing scatterplots like those above. For these types of questions, you will need to look at each scatterplot and determine which one has the dots in a configuration most similar to the one in the question. As stated in the previous problem, be sure that you know the difference between positive linear relationships and negative linear relationships for the exam.

The correct answer is D.

A negative linear relationship exists when an increase in one variable corresponds to a decrease in the other variable. This is represented by chart D.

23) The scatterplot provided below shows the relationship between the average hours of exercise each person had per day (H) and their strength ratings (S) on a scale of 1 to 5. Which of the following statements best describes the relationship between H and S?

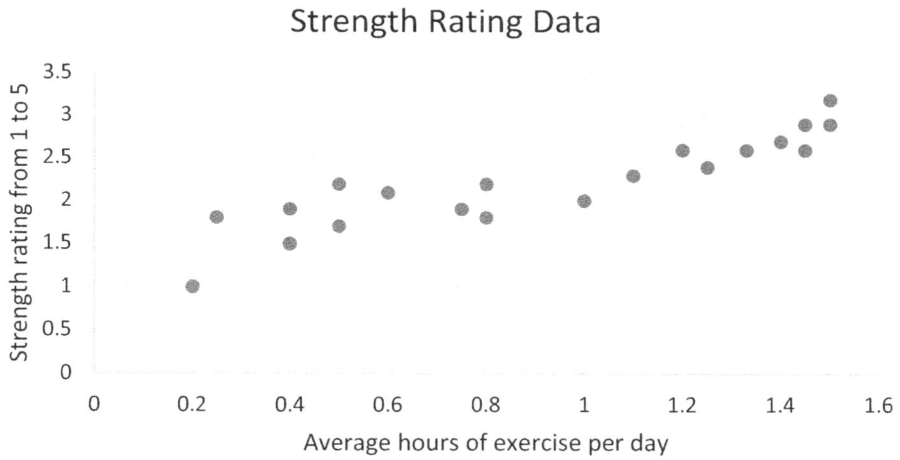

A) Increased amounts of exercise cause increases in performance ratings.
B) High strength ratings lead to increased amounts of exercise.
C) There is a positive linear relationship between H and S.
D) There is a negative linear relationship between H and S.
E) No relationship can be observed from the data.

This is a question on correlation on causation. For these types of questions, remember that correlation and causation are not the same. In other words, we cannot conclusively say that one variable causes another variable to happen when looking at data. We can only say that there is a correlation between the variables.

The correct answer is C.

From the chart, we can see that as the amount of exercise increases, the strength rating also increases. As mentioned previously, this is known as a positive linear relationship.

Be careful about answers A and B. They both are implying that there is a cause-and-effect relationship, rather than correctly identifying a correlation.

24) The pictograph below shows the number of pizzas sold in one day at a local pizzeria. Cheese pizzas sold for $10 each, pepperoni pizzas sold for $12, and the total sales of all three types of pizza was $310. What is the sales price of one vegetable pizza?

Cheese	▼ ▼ ▼
Pepperoni	▼ ▼
Vegetable	▼

Each ▼ represents 5 pizzas.

A) $5 B) $8 C) $9 D) $10 E) $12

This is an example of an exam question on interpreting data from pictographs. Each symbol on the pictograph represents a certain quantity of items, so remember to multiply by that amount in order to determine the totals for each group.

The correct answer is B.

First, determine how many cheese and pepperoni pizzas were sold. Each triangle symbol represents 5 pizzas.

Therefore, 15 cheese pizzas were sold:

3 symbols on the pictograph × 5 pizzas per symbol = 15 cheese pizzas

We also know that 10 pepperoni pizzas were sold:

2 symbols on the pictograph × 5 pizzas per symbol = 10 pepperoni pizzas

Then determine the value of these two types of pizzas based on the prices stated in the problem:

(15 cheese pizzas × $10 each) + (10 pepperoni pizzas × $12 each) =

$150 + $120 = $270

The remaining amount is allocable to the vegetable pizzas:

Total sales of $310 – $270 = $40 worth of vegetable pizzas

Since each triangle represents 5 pizzas, 5 vegetable pizzas were sold. We calculate the price of the vegetable pizzas as follows:

$40 worth of vegetable pizzas ÷ 5 vegetable pizzas sold = $8 per vegetable pizza

25) A zoo has reptiles, birds, quadrupeds, and fish. At the start of the year, they have a total of 1,500 creatures living in the zoo. The pie chart below shows percentages by category for the 1,500 creatures at the start of the year.

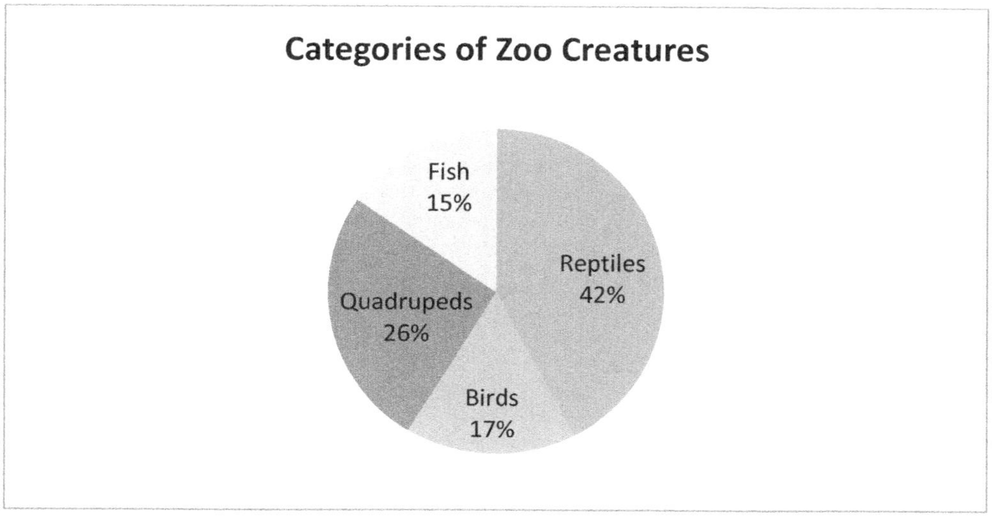

At the end of the year, the zoo still has 1,500 creatures, but reptiles constitute 40%, birds 23%, and quadrupeds 21%. How many more fish were there at the end of the year than at the beginning of the year?

A) 10 B) 11 C) 15 D) 16 E) 150

This question is asking you to interpret a pie chart that shows percentages by category. If you are asked to calculate changes to the data in the categories in the chart, be sure to multiply by the percentages at the beginning of the year and then do a separate calculation using the percentages at the end of the year.

The correct answer is C.

At the beginning of the year, 15% of the 1,500 creatures were fish, so there were 225 fish at the beginning of the year: 1,500 × 0.15 = 225

In order to find the percentage of fish at the end of the year, we first need to add up the percentages for the other animals: 40% + 23% + 21% = 84%

Then subtract this amount from 100% to get the remaining percentage for the fish: 100% – 84% = 16%

Multiply the percentage by the total to get the number of fish at the end of the year: 1,500 × 0.16 = 240

Then subtract the beginning of the year from the end of the year to calculate the increase in the number of fish: 240 – 225 = 15

26) The scatterplot below shows the relationship between the average number of visits to the recycling center per family per week for a given year and the amounts of garbage generated daily for these 25 families for the same year. For those families generating more than 1.5 pounds of garbage per day, what fraction had more than 2 trips to the recycling center per week?

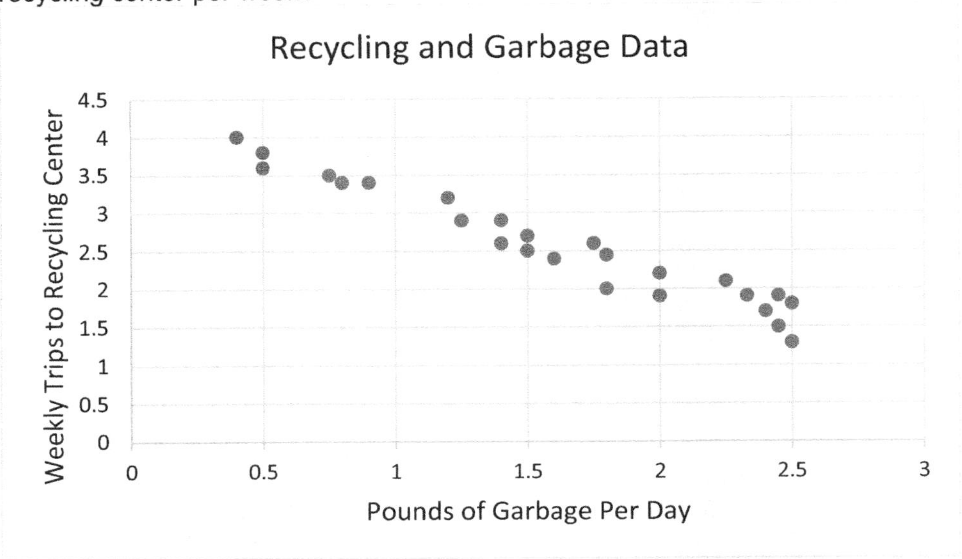

A) $\frac{1}{3}$ B) $\frac{1}{5}$ C) $\frac{6}{25}$ D) $\frac{7}{25}$ E) $\frac{8}{25}$

Read the facts carefully and then analyze the scatterplot. We need garbage amounts greater than 1.5 and recycling amounts more than 2. Be sure not to include the dots that are exactly 1.5 or 2, respectively, as the question is asking for amounts over these two figures.

The correct answer is B. There are five dots on the scatterplot that have garbage amounts greater than 1.5 and recycling amounts more than 2, as indicated in the rectangle in the scatterplot below. So, 5 out of 25 families meet the criteria for a fraction of $^5/_{25}$ or $^1/_5$.

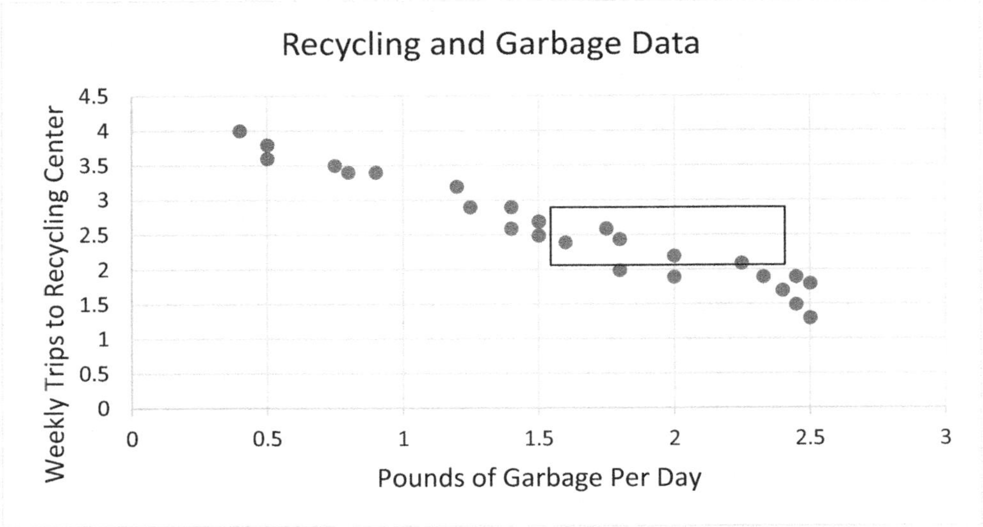

27) Use the information below to answer the question that follows.

• Classes in the morning last for 45 minutes, but classes in the afternoon last for 50 minutes. • Lunch begins promptly at 12:30 pm and finishes promptly at 1:00 pm. • There are 3 classes after lunch and 4 classes before lunch. • There are no breaks between classes or between classes and lunch.

Which one of the following statements could be true?
A) Classes begin at 9:30 am.
B) Classes begin at 10:00 m.
C) The second class after lunch begins at 2:00 pm.
D) The second class after lunch begins at 2:50 pm.
E) The last class of the day ends at 3:10 pm.

To answer these types of questions, read each of the answer choices one by one. Then compare each answer to the information provided to determine if that answer is true or false.

The correct answer is A.

If classes last for 45 minutes and there are 4 classes before lunch, the morning classes last for 3 hours in total: 45 minutes × 4 = 180 minutes = 3 hours

If lunch begins at 12:30, it is therefore possible for classes to begin three hours earlier at 9:30.

28) Which of the following methods of data representation would be best to represent the population of a certain city for each of the past ten years?
A) scatterplot B) pie chart C) bar graph D) 3-D illustration E) storyboard

For questions on data representation, remember the following: A scatterplot is used to see what kind of relationship exists between two items or events. A pie chart is useful to represent the amount or percentage of each group to the total of all groups. Bar graphs and line charts are usually used to represent changes to amounts over time. 3-D illustrations are useful for giving instructions on how to assemble something. Storyboards are visual representations used by graphic artists. They are not used for data.

The correct answer is C.

In the bar graph, you will have bars for all of the years, with each bar representing the amount of the population that year.

This would make it easy to see the changes to the population over time. A line graph would also be suitable, although that is not a choice for this question.

Statistics and Probability Problems:

29) The ages of 5 siblings are: 2, 5, 7, 12, and x. If the mean age of the 5 siblings is 8 years old, what is the age (x) of the 5th sibling?

A) 8 B) 10 C) 12 D) 14 E) 16

> This is a problem on determining the value that is missing from the calculation of a mean of a set of values. Remember that the mean is the same thing as the arithmetic average. In order to calculate the mean, you simply add up the values of all of the items in the set, and then divide by the number of items in the set. To solve problems like this one, set up an equation to calculate the mean, using x for the unknown value.

The correct answer is D.

Set up your equation to calculate the average, using x for the age of the 5th sibling:

$(2 + 5 + 7 + 12 + x) \div 5 = 8$

$(2 + 5 + 7 + 12 + x) \div 5 \times 5 = 8 \times 5$

$(2 + 5 + 7 + 12 + x) = 40$

$26 + x = 40$

$26 - 26 + x = 40 - 26$

$x = 14$

30) Members of a weight loss group report their individual weight loss to the group leader every week. During the week, the following amounts in pounds were reported: 1, 1, 3, 2, 4, 3, 1, 2, and 1. What is the mode of the weight loss for the group in pounds?

A) 1 B) 2 C) 3 D) 4 E) 18

> This is a question on mode. Mode is the value that occurs most frequently in a data set. For example, if 10 students scored 85 on a test, 6 students scored 90, and 4 students scored 80, the mode score is 85.

The correct answer is A.

The mode is the number that occurs the most frequently in the set.

Our data set is: 1, 1, 3, 2, 4, 3, 1, 2, 1.

The number 1 occurs 4 times in the set, which is more frequently than any other number in the set, so the mode is 1.

31) Mark's record of times for the 400-meter freestyle at swim meets this season is: 8.19, 7.59, 8.25, 7.35, 9.10. What is the median of his times?

A) 7.59 B) 8.19 C) 8.25 D) 8.096 E)40.48

> This question is asking you to find the median of a set of numbers. The median is the number that is in the middle of the set when the numbers are in ascending order.

The correct answer is B.

The problem provides the number set: 8.19, 7.59, 8.25, 7.35, 9.10

First of all, put the numbers in ascending order: 7.35, 7.59, 8.19, 8.25, 9.10

Then find the one that is in the middle: 7.35, 7.59, **8.19**, 8.25, 9.10

32) A student receives the following scores on her assignments during the term: 98.5, 85.5, 80.0, 97, 93, 92.5, 93, 87, 88, 82. What is the range of her scores?

A) 17.0 B) 18.0 C) 18.5 D) 89.65 E) 93.0

This is a question on calculating range. To calculate range, the lowest value in the data set is deducted from the highest value in the data set.

The correct answer is C.

To calculate the range, the low number in the set is deducted from the high number in the set.

The problem set is: 98.5, 85.5, 80.0, 97, 93, 92.5, 93, 87, 88, 82.

The high number is 98.5 and the low number is 80, so the range is: 98.5 – 80 = 18.5

33) An owner of a carnival attraction draws teddy bears out of a bag at random to give to prize winners. She has 10 brown teddy bears, 8 white teddy bears, 4 black teddy bears, and 2 pink teddy bears when she opens the attraction at the start of the day. The first prize winner of the day receives a brown teddy bear. What is the probability that the second prize winner will receive a pink teddy bear?

A) $^{1}/_{24}$ B) $^{1}/_{23}$ C) $^{2}/_{24}$ D) $^{2}/_{23}$ E) $^{1}/_{2}$

This is a question on calculating basic probability. First of all, calculate how many items there are in total in the data set, which is also called the "sample space" or (S). Then reduce the data set if further items are removed. Probability can be expressed as a fraction. The number of items available in the total data set at the time of the draw goes in the denominator. The chance of the desired outcome, which is also referred to as the event or (E), goes in the numerator of the fraction. You can determine the chance of the event by calculating how many items are available in the subset of the desired outcome.

The correct answer is D.

You need to determine the number of possible outcomes at the start of the day first of all.

The owner has 10 brown teddy bears, 8 white teddy bears, 4 black teddy bears, and 2 pink teddy bears when she opens the attraction at the start of the day. So, at the start of the day, she has 24 teddy bears: 10 + 8 + 4 + 2 = 24

Then you need to reduce this amount by the quantity of items that have been removed. The problem tells us that she has given out a brown teddy bear, so there are 23 teddy bears left in the sample space: 24 – 1 = 23

The event is the chance of the selection of a pink teddy bear. We know that there are two pink teddy bears left after the first prize winner receives their prize.

Finally, we need to put the event (the number representing the chance of the desired outcome) in the numerator and the number of possible remaining combinations (the sample space) in the denominator.

So, the answer is $^{2}/_{23}$.

34) A magician pulls colored scarves out of a hat at random. The hat contains 5 red scarves and 6 blue scarves. The other scarves in the hat are green. If a scarf is pulled out of the hat at random, the probability that the scarf is red is $^1/_3$. How many green scarves are in the hat?

A) 3 B) 4 C) 5 D) 6 E) 7

> This question is asking you to determine the value missing from a sample space when calculating basic probability. This is like other problems on basic probability, but we need to work backwards to find the missing value. First, set up an equation to find the total items in the sample space. Then subtract the quantities of the known subsets from the total in order to determine the missing value.

The correct answer is B.

First, we will use variable T as the total number of items in the set. The probability of getting a red scarf is $^1/_3$. So, set up an equation to find the total items in the data set.

$$\frac{5}{T} = \frac{1}{3}$$

$$\frac{5}{T} \times 3 = \frac{1}{3} \times 3$$

$$\frac{5}{T} \times 3 = 1$$

$$\frac{15}{T} = 1$$

$$\frac{15}{T} \times T = 1 \times T$$

$$15 = T$$

We have 5 red scarves, 6 blue scarves, and x green scarves in the data set that make up the total sample space, so now subtract the amount of red and blue scarves from the total in order to determine the number of green scarves.

$$5 + 6 + x = 15$$

$$11 + x = 15$$

$$11 - 11 + x = 15 - 11$$

$$x = 4$$

35) Becky rolls a fair pair of six-sided dice. One of the die is black and the other is red. Each die has values from 1 to 6. What is the probability that Becky will roll a 4 on the red die and a 5 on the black die?

A) $^1/_{36}$ B) $^2/_{36}$ C) $^1/_{12}$ D) $^2/_{12}$ E) $^{10}/_{12}$

> This is an advanced problem on understanding probability models. For these questions, you will usually have two items, like two dice or a coin and a die. Each item will have various outcomes, like heads or tails for the coin or the different numbers on the die. To solve problems like this one, it is usually best to write out the possible outcomes in a list. This will help you visualize the number of possible outcomes that make up the sample space. Then circle or highlight the events from the list to get your answer.

The correct answer is A.

In this case, we have two items, each of which has a variable outcome. There are 6 numbers on the black die and 6 numbers on the red die.

Using multiplication, we can see that there are 36 possible combinations: $6 \times 6 = 36$

To check your answer, you can list the possibilities of the various combinations:

(1,1)	(1,2)	(1,3)	(1,4)	(1,5)	(1,6)
(2,1)	(2,2)	(2,3)	(2,4)	(2,5)	(2,6)
(3,1)	(3,2)	(3,3)	(3,4)	(3,5)	(3,6)
(4,1)	(4,2)	(4,3)	(4,4)	(4,5)	(4,6)
(5,1)	(5,2)	(5,3)	**(5,4)**	(5,5)	(5,6)
(6,1)	(6,2)	(6,3)	(6,4)	(6,5)	(6,6)

If the number on the left in each set of parentheses represents the black die and the number on the right represents the red die, we can see that there is one chance that Becky will roll a 4 on the red die and a 5 on the black die.
The result is expressed as a fraction, with the event (chance of the desired outcome) in the numerator and the total sample space (total data set) in the denominator.

So, the answer is $^{1}/_{36}$.

36) The school board wants to poll a sample of students to get their opinions on dropping the music program in favor of having more sports programs. Which one of the following methods will result in the most statistically valid information about the opinions of all of the students at the high school?

A) To select ten students at random from each grade at the school
B) To speak to all of the members of the high school football team
C) To ask two members of each grade at random as they leave band practice
D) To give questionnaires out to the freshmen and sophomore students
E) To gather the opinions from students who are willing to speak to the school board

This question is asking you about how best to use random sampling to draw conclusions about data. For information to be statistically valid, the data must be taken at random from a sample set of respondents that best represents the entire group.

The correct answer is A.

Our entire group in this problem is all of the students at the high school. So, it would be best to select ten students at random from each grade at the school.

The other answer choices would be biased in favor of members of certain groups, namely football players (answer B), band participants (answer C), younger students (answer D), and students who are not afraid to speak to the school board (answer E).

37) Which of the following are statistical questions? You may select more than one answer.

_____ (1) How long is that piece of string?

_____ (2) How many customers go to that coffee shop on Saturdays?

_____ (3) What time is history class?

_____ (4) How do students rate Mrs. Brown's calculus class?

_____ (5) What are the perimeter measurements of the high school parking lot?

Statistical questions usually ask about opinions and behaviors. In addition, statistical questions can have a variety of different answers. On the other hand, questions about measuring time and distance are non-statistical questions because they have only one possible answer.

The correct answers are (2) and (4).

“How many customers go to that coffee shop on Saturdays?” is a statistical question because the number of customers in the shop can be counted. The number will also vary every Saturday.

“How do students rate Mrs. Brown’s calculus class?” is also a statistical question because some students will have positive opinions, others will have negative opinions, and still others will have mixed opinions. In addition, the various ratings can be counted.

Algebra Problems:

38) If $\frac{3}{4}x - 2 = 4$, x = ?

A) $\frac{8}{3}$ B) $\frac{1}{8}$ C) 8 D) –8 E) 24

This is a problem requiring you to solve an expression that contains a single variable, a fraction, and integers. First, isolate the integers and then eliminate the fraction. Finally, divide to find the value of the variable.

The correct answer is C.

Isolate the integers to one side of the equation.

$$\frac{3}{4}x - 2 = 4$$

$$\frac{3}{4}x - 2 + 2 = 4 + 2$$

$$\frac{3}{4}x = 6$$

Then get rid of the fraction by multiplying both sides by the denominator.

$$\frac{3}{4}x \times 4 = 6 \times 4$$

$$3x = 24$$

Then divide to solve the problem.

$$3x \div 3 = 24 \div 3$$

$$x = 8$$

39) If $2(3x - 1) = 4(x + 1) - 3$, what is the value of x? State your answer as a fraction in the spaces provided.

Numerator value of x = ___________

Denominator value of x = ___________

This problem requires you to solve an algebraic expression that contains one variable (x) on both sides of the equation. When the variable is used on both sides of the equation, you should perform the multiplication on the parentheticals first. Then isolate x to solve the problem.

The correct answers are 3 (numerator value) and 2 (denominator value).

Perform the multiplication on the terms in the parentheses.
$2(3x - 1) = 4(x + 1) - 3$
$6x - 2 = (4x + 4) - 3$

Then simplify.
$6x - 2 = 4x + 4 - 3$
$6x - 2 = 4x + 1$
$6x - 2 - 1 = 4x + 1 - 1$
$6x - 3 = 4x$

Then isolate x to get your answer.
$6x - 3 = 4x$
$6x - 4x - 3 = 4x - 4x$
$2x - 3 = 0$
$2x - 3 + 3 = 0 + 3$
$2x = 3$
$2x \div 2 = 3 \div 2$
$x = {}^{3}/_{2}$

40) In a group of 150 athletes, 75 are marathon runners, as represented by circle M, and 45 are triathletes, as represented by circle T. If 25 athletes are both marathon runners and triathletes, how many of the athletes are neither marathon runners nor triathletes?

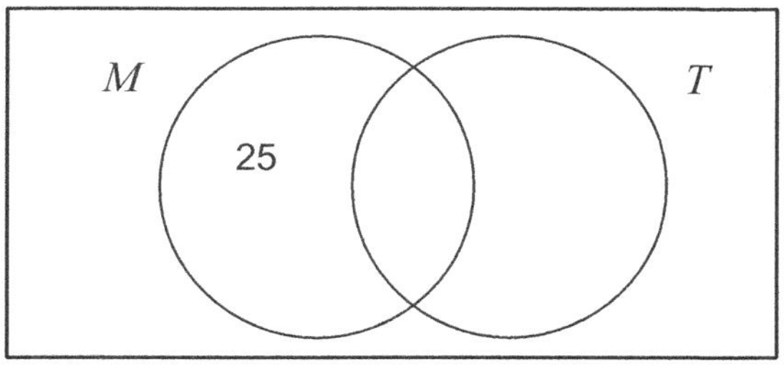

A) 30 B) 45 C) 55 D) 70 E) 95

In a Venn diagram, the number of items in the intersection where the circles overlap is included in both groups. Be sure not to double count these items.

The correct answer is C.

First, add the number of marathon runners and triathletes together: $75 + 45 = 120$

Then subtract 25 from this amount as it would have been double-counted since these athletes were in both groups: $120 - 25 = 95$

Then subtract this from the total to determine how many other athletes there are:
$150 - 95 = 55$

41) Which of the following equations is equivalent to $\frac{x}{5} + \frac{y}{2}$?

A) $\frac{x+y}{7}$ B) $\frac{2x+5y}{10}$ C) $\frac{5x+2y}{10}$ D) $\frac{2x+5y}{7}$ E) $\frac{5y}{2x}$

> This problem is asking you to find an equivalent expression for a mathematical equation that contains fractions. To add fractions, find the lowest common denominator first and then add the numerators.

The correct answer is B.

You need to find the lowest common denominator. Then add the numerators together as shown.

$$\frac{x}{5} + \frac{y}{2} =$$

$$\left(\frac{x}{5} \times \frac{2}{2}\right) + \left(\frac{y}{2} \times \frac{5}{5}\right) =$$

$$\frac{2x}{10} + \frac{5y}{10} =$$

$$\frac{2x+5y}{10}$$

42) Which of the following steps will solve the equation for x: $4x - 3 = 2$

A) Add 3 to both sides of the equation, and then divide both sides by 4.
B) Add 3 to both sides of the equation, and then subtract 4 from both sides.
C) Add 2 to both sides of the equation, and then divide both sides by 4.
D) Subtract 2 from both sides of the equation, and then divide both sides by –3.
E) Divide both sides of the equations by 4, and then subtract 3 from both sides.

> For this question, you need to find the answer that provides a narrative explanation for the solution to the problem, rather than calculating the solution itself.

The correct answer is A.

Remember that to solve problems like this, you need to deal with the integers (whole numbers) and then isolate the variable (x).

The solution is as follows:

$4x - 3 = 2$

$4x - 3 + 3 = 2 + 3$ (Add 3 to both sides of the equation.)

$4x = 5$

$4x \div 4 = 5 \div 4$ (Divide both sides by 4.)

$x = \frac{5}{4}$

43) Toby is going to buy a car. The total purchase price of the car is represented by variable C. He will pay D dollars immediately, and then he will make equal payments (P) each month for a certain number of months (M). Which equation below represents the amount of his monthly payment (P)?

A) $\frac{C-D}{M}$ B) $\frac{C}{M} - D$ C) $\frac{M}{C-D}$ D) $D - \frac{C}{M}$ E) $\frac{C}{M}$

This problem requires you to set up an algebraic equation based on facts in a real-life problem. In this problem, we need to calculate a monthly payment after a down payment has been made. Deduct the down payment from the purchase price, and then divide by the number of months to solve the problem.

The correct answer is A.

The total amount that Toby has to pay is represented by C.

He is paying D dollars immediately, so we can determine the remaining amount that he owes by deducting his down payment from the total.

So, the remaining amount owing is represented by the equation: C – D

We have to divide the remaining amount owing by the number of months (M) to get the monthly payment (P): P = (C – D) ÷ M = $\frac{C-D}{M}$

44) Fatima drove into town at a rate of 50 miles per hour. She shopped in town for 20 minutes, and then drove home on the same route at a rate of 60 miles per hour. Which of the following equations best expresses the total time (Tt) that it took Fatima to make the journey and do the shopping? Note that the variable D represents the distance in miles from Fatima's house to town.

A) $Tt + 20 \text{ minutes} = 110 \times D$
B) $Tt + 20 \text{ minutes} = [(50 + 60) \div 2] \times D$
C) $Tt = [(D \div 50) + (D \div 60)] + 20 \text{ minutes}$
D) $Tt = D \div 110$
E) $Tt = (D \div 110) + 20 \text{ minutes}$

This problem asks you to set up an algebraic equation based on facts in a real-life problem. In this problem, we need to calculate the time spent on a journey. Read the problem carefully and make sure that all of the required facts are represented in your final equation.

The correct answer is C.

The amount of time in hours (T) multiplied by miles per hour (mph) gives us the distance traveled (D). So, the equation for distance traveled is: $T \times \text{mph} = D$

The problem tells us that we need to calculate T, so we need to isolate T by changing our equation as follows:

$T \times \text{mph} = D$

$(T \times \text{mph}) \div \text{mph} = D \div \text{mph}$

$T = D \div \text{mph}$

In our problem, Fatima drives home on the same route that she took into town, so we need to calculate the traveling time for the journey into town, as well as for the journey home: $(D \div 50) + (D \div 60) = T$

Then add back the 20 minutes she spent in town to get the total time:

$Tt = [(D \div 50) + (D \div 60)] + 20 \text{ minutes}$

45) A baseball team sells T-shirts and sweatpants to the public for a fundraising event. The total amount of money the team earned from these sales was \$850. Variable t represents the number of T-shirts sold and variable s represents the number of sweatpants sold. The total sales in dollars is represented by the equation $25t + 30s$. The amount earned by selling sweatpants is what fraction of the total amount earned?

A) $s/850$ B) $30s/850$ C) $(25t + 30s)/850$ D) $t/850$ E) $25t/850$

This problem asks you to express part of an algebraic equation as a fraction of the entire equation. You can sometimes substitute a numerical or dollar value in the expression, as in this problem.

The correct answer is B.

We need to set up a fraction, the numerator of which consists of the amount of sales in dollars for sweatpants, and the denominator of which consists of the total amount of sales in dollars for both items.

The problem tells us that the amount of sales in dollars for sweatpants is $30s$ and the total amount of sales is \$850, so the answer is $30s/850$.

46) The information provided in the box below describes three locations on a map. Use the information provided to answer the question that follows.

• The police station is 10 miles away from the fire station • The fire station is 6 miles away from the hospital.

Based on the information in the box, what conclusions can be made?

A) The police station is no more than 6 miles away from the hospital.
B) The police station is no more than 10 miles away from the hospital.
C) The police station is exactly 6 miles away from the hospital.
D) The police station is no more than 16 miles away from the hospital.
E) No conclusions may be made since essential information is missing.

Read the facts carefully for questions involving maps and then make conclusions based on the information provided. You may find it helpful to draw diagrams to help you answer these types of questions. For some questions, a diagram may be provided.

The correct answer is D.

For questions about distance like this one, keep in mind that the locations may or may not lie on a straight line. In the following layout, the police station would be 16 miles from the hospital if the locations were laid out on the map like this:

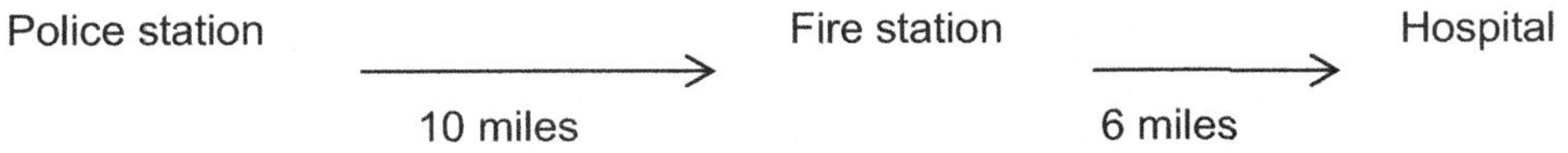

However, the locations could also be laid out like this:

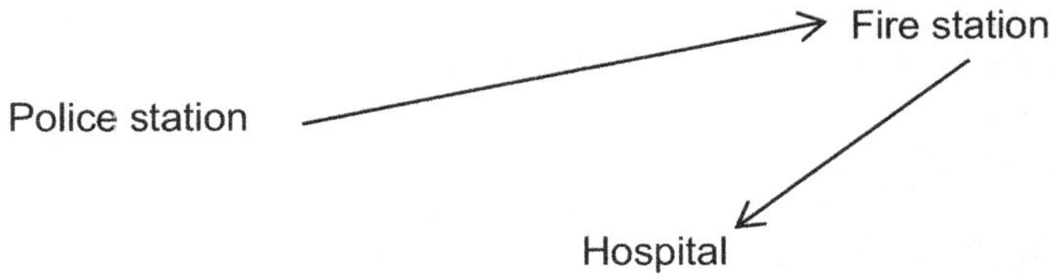

We can see that the locations will be the farthest from each other if they are laid out on a straight line as in the first example above.

In other words, a person could always go to the hospital by traveling to the fire station from the police station (10 miles) and then traveling from the fire station to the hospital (6 miles).

Therefore, the police station would never be more than 16 miles away from the hospital, regardless of the layout.

47) The following three locations are on a map of an amusement park: Ferris wheel, roller coaster, and entrance. On the map, which is drawn to scale, the distance between the Ferris wheel and the entrance is 8 inches, and the distance between the entrance and the roller coaster is 6.5 inches. The actual distance between the Ferris wheel and the entrance is 350 yards. What is the actual distance between the entrance and the roller coaster to the nearest yard?
A) 294 B) 284 C) 350 D) 360 E) 43.75

For questions about scale, divide the actual distance by the corresponding distance on the map to get the scale of the map.

The correct answer is B.

The actual distance between the Ferris wheel and the entrance is 350 yards, and this is represented by 8 inches on the map, so divide to get the scale: $350 \div 8 = 43.75$

In other words, every inch on the map represents 43.75 yards.

Now multiply the scale of the map by the number of inches between the entrance and the roller coaster to get the actual distance: $6.5 \times 43.75 = 284.375$

Finally, round to 284 yards.

48) Indicate which of the following statements are true for all positive numbers x, y and z. You may choose more than one answer.
A) $x \times (y + z) = (y + z) \times x$
B) $x - (y + z) = x - y + z$
C) $x \times (y + z) = (x \times y) + z$
D) $x + (y - z) = (x + y) - z$

Consider the communicative, associate, and distributive properties when you are analyzing each of the answer choices.

The correct answers are A and D.

Answer A is true because of the communicative property of multiplication. In other words, $x \times y = y \times x$

Answer D is true because y is a positive number and z, as a negative number, is the last operation to be performed on both sides of the equation.

Geometry Problems:

49) The figure shown below consists of five equal squares. The perimeter of the entire figure is 108 feet. What is the area the entire figure in square feet? [Area of a square or rectangle = Length × Width; Perimeter = (2 × Length) + (2 × Width)]

A) 45 B) 81 C) 324 D) 405 E) 540

Three sides of each of the exterior squares form the cross, so the cross has twelve sides. Get the length of one side of each square first. Then find the area of each square. Then multiply by the number of squares that makeup the figure.

The correct answer is D.

The figure has 12 sides, so divide 108 by 12 to get the length of one side: $108 \div 12 = 9$

Then determine the area of each square. Remember that the area of a square or rectangle is length × width: $9 \times 9 = 81$

The figure consists of five squares, so multiply to solve: $81 \times 5 = 405$ square feet

50) A small circle has a radius of 5 inches, and a larger circle has a radius of 8 inches. What is the difference in inches between the circumferences of the two circles?
A) 3 B) 6 C) 6π D) 9π E) 39π

The question above is asking you to calculate the circumference of two circles. The formula for the circumference of a circle is: Circumference of a circle = $2\pi R$ = ($2 \times \pi \times$ radius) Radius = diameter $\times \frac{1}{2}$

The correct answer is C.

Use the formula from above: ($2 \times \pi \times$ radius)

So, we calculate the circumference of the large circle as: $2 \times \pi \times 8 = 16\pi$

The circumference of the small circle is: $2 \times \pi \times 5 = 10\pi$

Then, we subtract to get our solution: $16\pi - 10\pi = 6\pi$

51) Which of the following statements about isosceles triangles is true?
A) Isosceles triangles have two equal sides.
B) When an altitude is drawn in an isosceles triangle, two equilateral triangles are formed.
C) The base of an isosceles triangle must be shorter than the length of each of the other two sides.
D) The sum of the measurements of the interior angles of an isosceles triangle must be equal to 360°.
E) If two sides of an isosceles triangle are equal, the angles opposite them are right triangles.

This question assesses your knowledge of the rules for triangles and angles.

Remember these principles on angles and triangles for your exam:

The sum of all three angles in any triangle must be equal to 180 degrees.

An isosceles triangle has two equal sides and two equal angles.

An equilateral triangle has three equal sides and three equal angles.

Angles that have the same measurement in degrees are called congruent angles.

Equilateral triangles are sometimes called congruent triangles.

Two angles are supplementary if they add up to 180 degrees. This means that when the two angles are placed together, they will form a straight line on one side.

Two angles are complementary (sometimes called adjacent angles) if they add up to 90 degrees. This means that the two angles will form a right triangle.

A parallelogram is a four-sided figure in which opposite sides are parallel and equal in length. Each angle will have the same measurement as the angle opposite to it, so a parallelogram has two pairs of opposite angles.

The correct answer is A.

As stated above, an isosceles triangle has two equal sides, so answer A is correct.

If an altitude is drawn in an isosceles triangle, we have to put a straight line down the middle of the triangle from the peak to the base. Dividing the triangle in this way would form two right triangles, rather than two equilateral triangles. So, answer B is incorrect.

The base of an isosceles triangle can be longer than the length of each of the other two sides, so answer C is incorrect.

The sum of all three angles of any triangle must be 180 degrees, rather than 360 degrees. So, answer D is incorrect.

By definition a triangle must have three sides. Also remember that all three angles inside the triangle must add up to 180 degrees and that right angles measure 90 degrees.

Therefore, the angles opposite the two equal sides of an isosceles triangle cannot be right triangles because $2 \times 90^\circ = 180^\circ$. In this case, there would be no room for the third angle. So, answer E is incorrect.

Go on to the next page.

52) Which of the following dimensions would be needed in order to find the area of the figure?

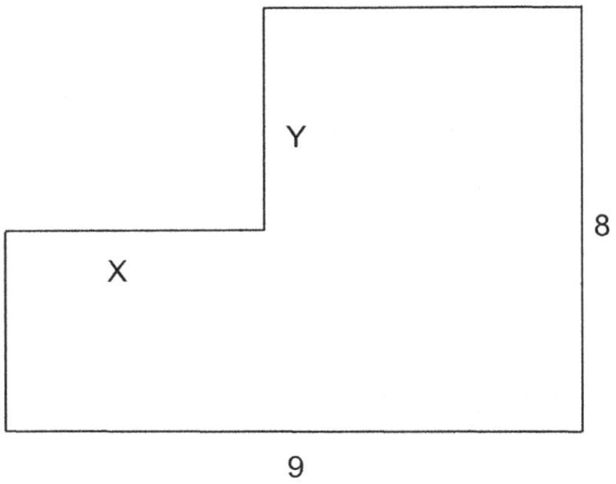

A) X only B) Y only C) Both X and Y D) Either X or Y E) Neither X nor Y

This question is asking you to calculate the area of a hybrid shape. To solve problems like this one, try to visualize two rectangles. The first rectangle would measure 8 × 9 and the second rectangle would measure X × Y.

The correct answer is C.

Essentially a rectangle is missing at the upper left-hand corner of the figure.

We would need to know both the length and width of the "missing" rectangle in order to calculate the area of our figure.

So, we need to know both X and Y in order to solve the problem.

53) The area of a square is 64 square units. This square is made up of smaller squares that measure 4 square units each. How many of the smaller squares are needed to make up the larger square?
A) 8 B) 12 C) 16 D) 24 E) 32

This question is asking you to determine the relationships between square figures of different sizes. For problems about placing small squares inside a larger square, you can simply divide the size of the smaller squares into the size of the larger square in order to determine how many small squares are required.

The correct answer is C.

We simply divide to get the answer: 64 ÷ 4 = 16

54) The illustration below shows an isosceles triangle. The entire triangle has a base of 9 and a height of 18.

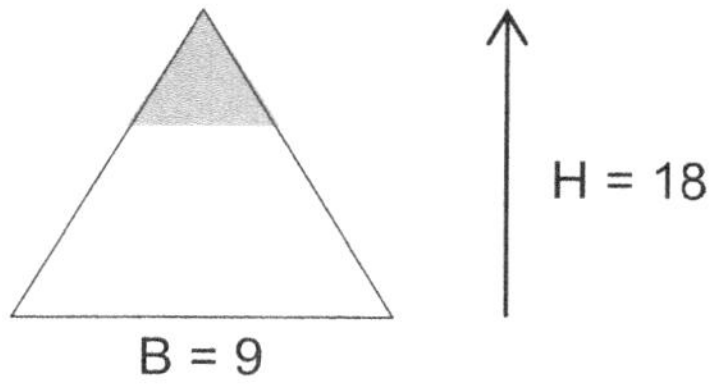

The shaded portion at the top of the triangle has a height of 6. What fraction expresses the area of the shaded portion to the area of the entire triangle?

State your answer as a fraction in the spaces provided.

Numerator value = ___________

Denominator value = __________

> This question is asking you to calculate the area of a triangle.
>
> The formula for the area of a triangle is: $\frac{B \times H}{2}$
>
> This can also be expressed as: (base × height) ÷ 2

The correct answers are a numerator of 9 and a denominator of 81.

Another possible answer is a numerator of 1 and a denominator of 9.

First, we need to calculate the area of the entire triangle: $(9 \times 18) \div 2 = 81$

Then, we need to calculate the base of the shaded portion.

Since the height of the shaded portion is 6 and the height of the entire triangle is 18, we know by using the rules of similarity that the ratio of the base of the shaded portion to the base of the entire cone is $^6/_{18}$ or $^1/_3$.

Using this fraction, we can calculate the base for the shaded portion.

The base of the entire triangle is 9, so the base of the shaded portion is 3: $9 \times {}^1/_3 = 3$

Then, calculate the area of the shaded portion: $(3 \times 6) \div 2 = 9$

So, we express the volume of the shaded portion to the volume of the entire cone as: $^9/_{81}$

Alternatively, we can simplify this to $^1/_9$.

55) In the figure below, lines 1 and 2 are parallel. What is the measurement in degrees of angle y?

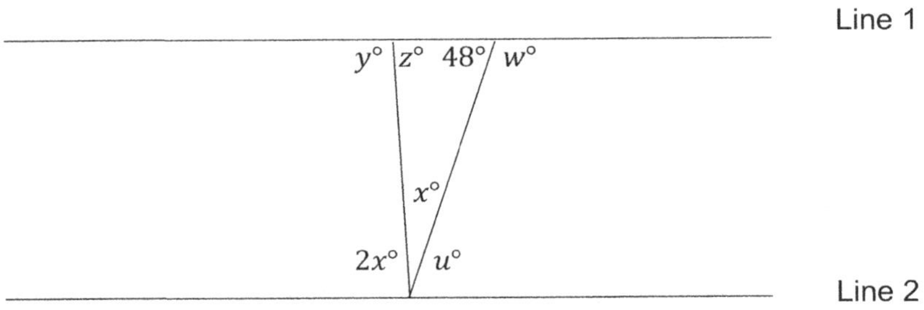

A) 48 B) 44 C) 88 D) 92 E) 132

When two parallel lines are cut by a transversal (a straight line that runs through or connects both of the parallel lines), opposite angles are formed. These opposite angles are equal in measure. In the illustration above, angle z and the angle that measures $2x°$ are opposite angles. Angle u and the 48° angle are also opposite angles.

The correct answer is D.

Because of the principles stated above, angle z measures $2x°$.

Remember that there are 180° degrees in total when we add up the three angles of any triangle and that two angles that form a straight line also add up to 180°.

So, we set up our equation for the triangle as follows: $180° = z° + x° + 48°$

We simplify this as shown:

$180° = z° + x° + 48°$
$180° = 2x° + x° + 48°$
$180° = 3x° + 48°$
$180° - 48° = 3x°$
$132° = 3x°$
$44° = x$

Then calculate the measurement of angle z:

$z° = 2x°$
$z° = 2 \times 44°$
$z = 88$

Finally, since $y° + z°$ forms a straight line and because straight lines measure 180°, we can calculate the measurement of angle y:

180° (for the straight line) – 88° (for angle z) = 92°

56) Consider rectangular figure WXYZ below:

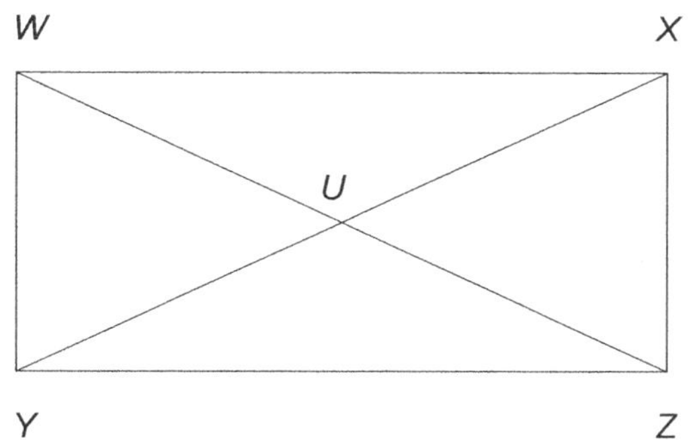

Which of the following must be true?

A) $\angle YWX \cong \angle YWU$

B) $\angle YUW \cong \angle WUX$

C) $\overline{WY} \cong \overline{UY}$

D) $\overline{WY} \cong \overline{WX}$

E) $\overline{WY} \cong \overline{XZ}$

> Remember that opposite sides of a rectangle are equal in length. You may also wish to refer back to the rules on angles provided previously.

The correct answer is E.

$\overline{WY}$ and $\overline{XZ}$ are opposite sides of the rectangle, and opposite sides are equal in length.

The two opposite sides of a rectangle are also called corresponding sides.

Praxis Core Math Practice Set 2

For practice tests 2 and 3, you should refer to the formula sheet in the appendix of the book when needed.

1) Which of the following is the greatest?
A) 0.540
B) 0.054
C) 0.045
D) 0.5045
E) 0.0054

2) Brooke wants to put new flooring in her living room. She will buy the flooring in square pieces that measure 1 square foot each. The entire room is 12 feet by 16 feet. The bookcases are one and a half feet deep from front to back. Flooring will not be put under the bookcases. Each piece of flooring costs $4.25. A diagram of her living room is provided.

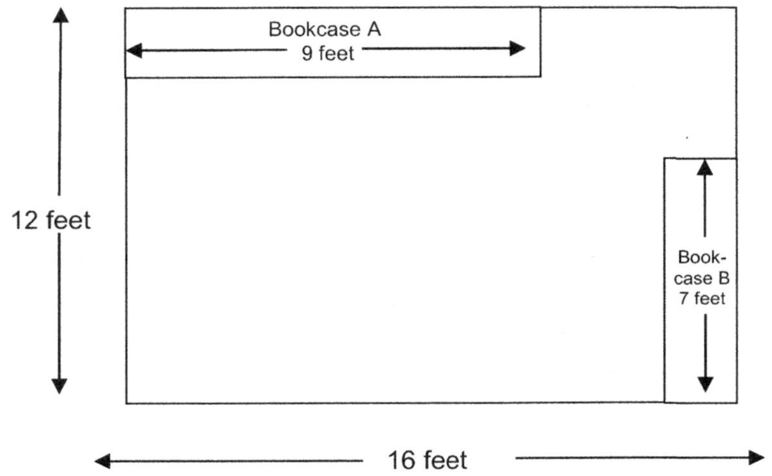

How much will Brooke pay to cover her living room floor? Provide the cost to the nearest dollar.
A) 168
B) 192
C) 714
D) 758
E) 816

3) Farmer Brown has a field in which cows craze. He is going to buy fence panels to put up a fence along one side of the field. Each panel is 8 feet 6 inches long. He needs 11 panels to cover the entire side of the field. How long is the field?
A) 60 feet 6 inches
B) 72 feet 8 inches
C) 93 feet 6 inches
D) 102 feet 8 inches
E) 110 feet 6 inches

4) In the Venn diagram below, circle *A* represents the integers from 3 to 13 inclusive, and circle B represents the integers 5 to 15 inclusive. How many integers are represented in region C of the diagram?

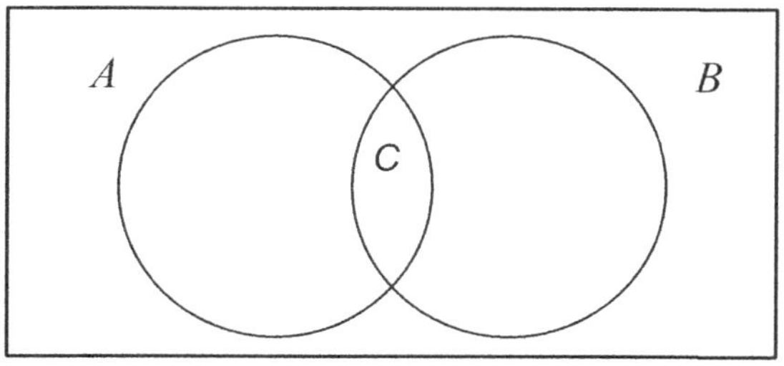

A) 2
B) 8
C) 9
D) 10
E) 11

5) If the value of x is between 0.0007 and 0.0021, which of the following could be x?
A) 0.0012
B) 0.0006
C) 0.0022
D) 0.022
E) 0.08

6) The total funds, represented by variable F, available for P charity projects is represented by the equation F = $500P + $3,700. If the charity has $40,000 available for projects, what is the greatest number of projects that can be completed?
A) 72
B) 73
C) 74
D) 79
E) 80

Go on to the next page.

7) The students at Lyndon High School have been asked about their plans to attend the Homecoming Dance. The chart below shows the responses of each grade level by percentages. Which figure below best approximates the percentage of the total number of students from all four grades who will attend the dance? Note that each grade level has roughly the same number of students.

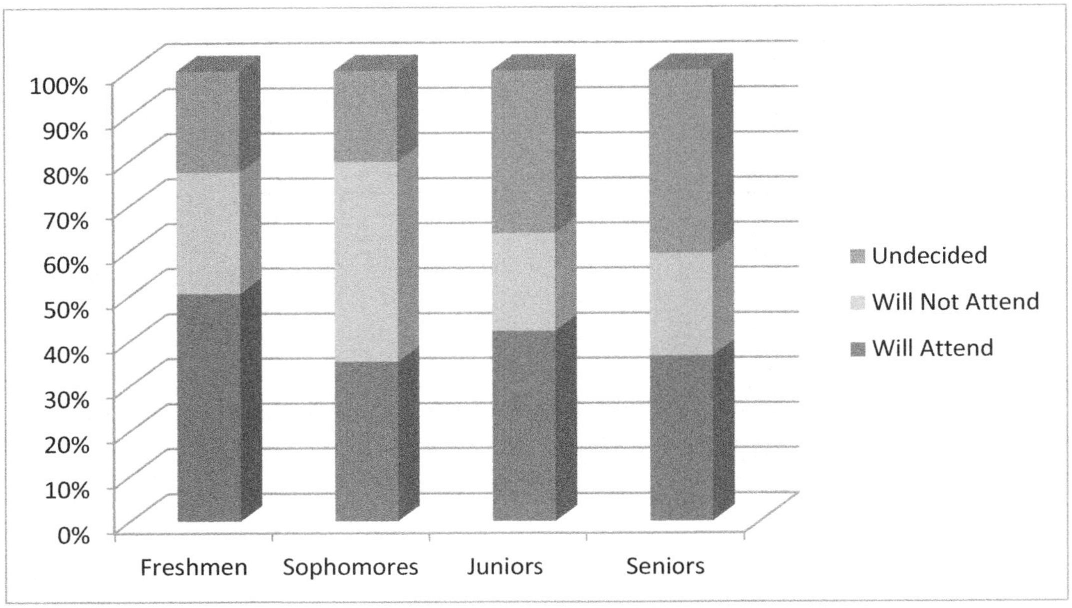

A) 25%
B) 35%
C) 45%
D) 55%
E) 60%

8) Look at the table below and answer the question that follows.

Sunday	Monday	Tuesday	Wednesday	Thursday	Friday	Saturday
–10°F	–9°F	1°F	6°F	8°F	13°F	12°F

What was the median temperature for the week?
A) 1°
B) 3°
C) 6°
D) 22°
E) 24°

9) During each flight, a flight attendant must count the number of passengers on board the aircraft. The morning flight had 52 passengers more than the evening flight, and there were 540 passengers in total on the two flights that day. How many passengers were there on the evening flight?
A) 244
B) 296
C) 488

D) 540
E) 592

10) A cafeteria serves spaghetti to senior citizens on Fridays. The spaghetti comes prepared in large containers, and each container holds 15 servings of spaghetti. The cafeteria is expecting 82 senior citizens this Friday. What is the least number of containers of spaghetti that the cafeteria will need in order to serve all 82 people?
A) 4
B) 5
C) 6
D) 7
E) 15

11) A caterpillar travels 10.5 inches in 45 seconds. How far will it travel in 6 minutes?
A) 45 inches
B) 63 inches
C) 64 inches
D) 84 inches
E) 90 inches

12) Which one of the values will correctly satisfy the following mathematical statement: $^{2}/_{3}$ < ? < $^{7}/_{9}$
A) $^{1}/_{3}$
B) $^{1}/_{5}$
C) $^{2}/_{6}$
D) $^{1}/_{2}$
E) $^{7}/_{10}$

13) Determine the ratio of side XZ to side WX in the figure below. Note that ΔUWX and ΔXYZ are similar.

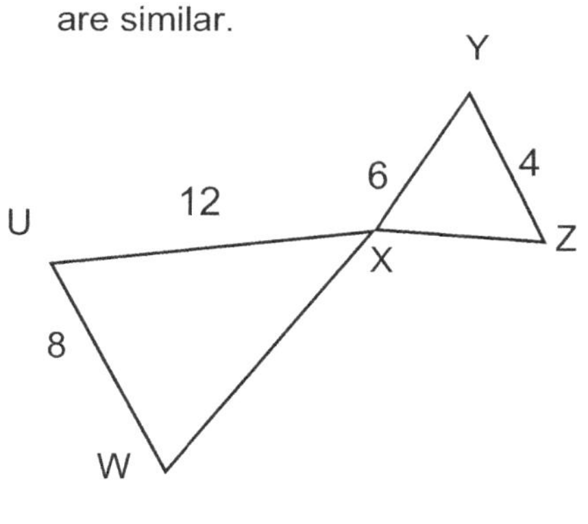

A) $\frac{3}{2}$

B) $\frac{2}{3}$

C) $\frac{4}{9}$

D) $\frac{9}{4}$

E) $\frac{1}{2}$

14) Which of the following are equivalent to 120,000,000? You may choose more than one answer.
A) 120 thousand thousands
B) 12 thousand thousands
C) 120 million
D) 120 billion

15) A company is making its budget for the cost of employees to attend conferences for the year. It costs \$7,500 per year in total for the company plus C dollars per employee. During the year, the company has E employees. If the company has budgeted \$65,000 for conference attendance, which equation can be used to calculate the maximum cost per employee?
A) (\$65,000 – \$7,500) ÷ E
B) (\$65,000 – \$7,500) ÷ C
C) (C – \$7,500) ÷ E
D) \$65,000 ÷ E
E) (\$65,000 ÷ E) – \$7,500

16) The pictograph below illustrates the results of a customer satisfaction survey by region. Each of the four regions has one salesperson. Salespeople in each region receive bonuses based on the amount of positive customer feedback they receive. If the salespeople from all four regions received \$540 in bonuses in total, how much bonus money does the company pay each individual salesperson per satisfied customer?

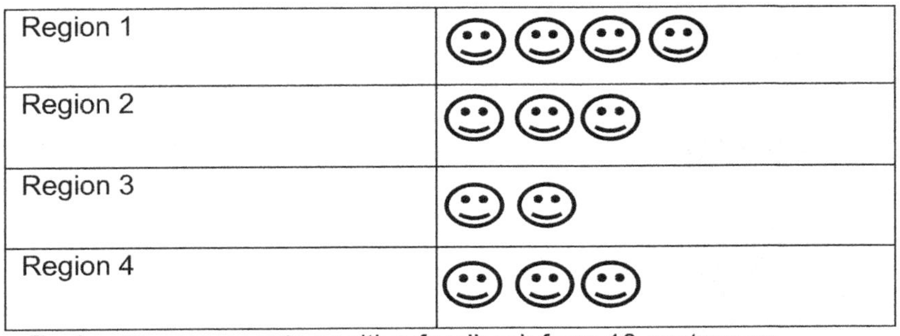

Each  represents positive feedback from 10 customers.

A) \$4.00
B) \$4.50
C) \$4.90
D) \$5.00
E) \$5.40

17) A megastore gives away a cell phone for free to any customer who spends \$2,250 or more in the store during the month of July. The store had \$50,250 in total sales income for July, and 310 customers made purchases in the store during that month. Which equation below can be used to calculate the number of free cell phones that the store gave away during the month of July?
A) (\$50,250 ÷ \$2,250)
B) (\$50,250 ÷ 310)
C) (\$50,250 ÷ \$2,250) ÷ 310
D) (\$50,250 – \$2,250) ÷ 310
E) Cannot be determined from the information provided.

18) Which of the following is equivalent to the expression $2(x + 2)(x - 3)$ for all values of x?

A) $2x^2 - 2x - 12$
B) $2x^2 - 10x - 6$
C) $2x^2 + 2x - 12$
D) $2x^2 + 10x - 6$
E) $2x^2 - x - 3$

19) A plumber charges $100 per job, plus $25 per hour worked. She is going to do 5 jobs this month. She will earn a total of $4,000. How many hours will she work this month? Please write your answer in the space provided.

__________ hours this month

20) What are two possible values of x for the following equation? $x^2 + 6x + 8 = 0$

A) 1 and 2
B) 2 and 4
C) 6 and 8
D) –2 and –4
E) –3 and –4

21) The ratio of teachers to students on a class trip is 2 to 6. Which of the following could possibly be the total number of teachers and students on the trip?

A) 90
B) 92
C) 94
D) 96
E) 98

22) A is 3 times B, and B is 3 more than 6 times C. Which of the following describes the relationship between A and C?

A) A is 9 more than 18 times C.
B) A is 3 more than 3 times C.
C) A is 3 more than 18 times C.
D) A is 6 more than 3 times C.
E) A is 18 more than 9 times C.

Go on to the next page.

23) The graph below shows the relationship between the number of days of rain per month and the amount of people who exercise outdoors per month. What relationship can be observed?

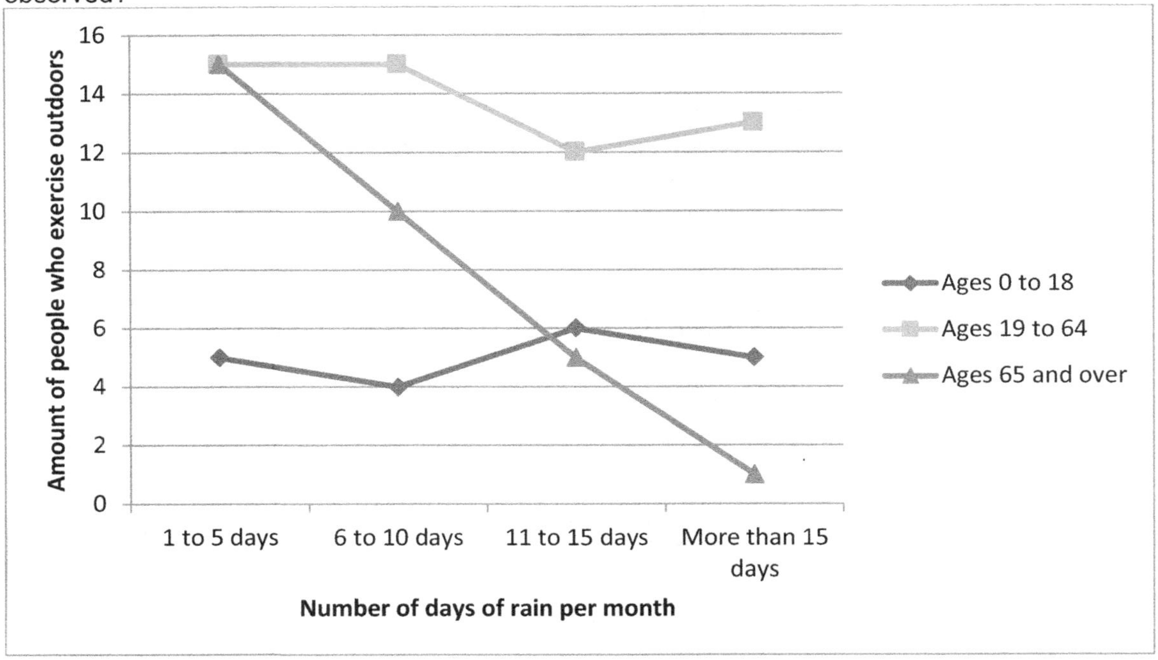

A) Young children are reliant upon an adult in order to exercise outdoors.
B) The exercise habits of working age people seem to fluctuate proportionately to the amount of rainfall.
C) In the 19 to 64 age group, there is a negative relationship between the number of days of rain and the amount of people who exercise outdoors.
D) People aged 65 and over seem less inclined to exercise outdoors when there is more rain.
E) No relationship can be observed because of the disparities inherent among the age groups.

24) A pharmacist works at a constant rate, filling 3 prescriptions every twenty minutes. How long will it take her to fill 45 prescriptions?
A) 1 hour 45 minutes
B) 2 hours 45 minutes
C) 4 hours 45 minutes
D) 5 hours
E) 6 hours 45 minutes

25) Consider the scatterplot below and then choose the best answer from the options that follow.

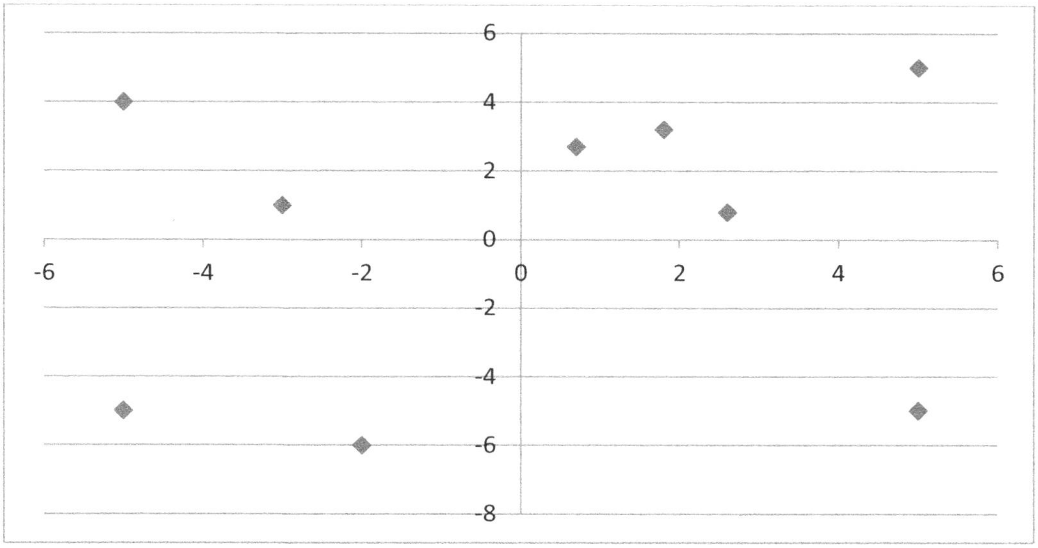

A) The scatterplot suggests a strong positive linear relationship between x and y.
B) The scatterplot suggests a strong negative linear relationship between x and y.
C) The scatterplot suggests a weak positive linear relationship between x and y.
D) The scatterplot suggests a weak negative linear relationship between x and y.
E) The scatterplot suggests that there is no relationship between x and y.

26) The graph of a line is shown on the xy plane below. The point that has the y-coordinate of 45 is not shown. What is the corresponding x-coordinate of that point?

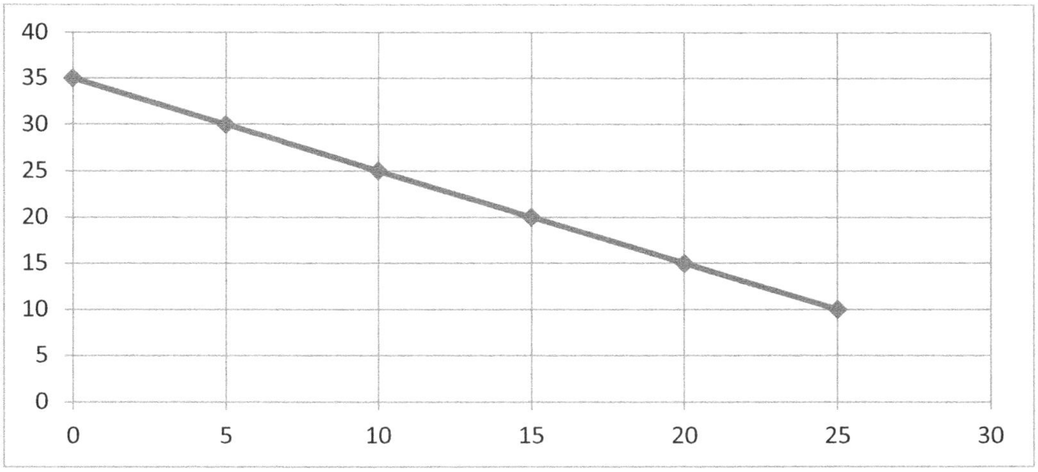

A) –10
B) –5
C) 0
D) 5
E) 30

27) Find the equivalent for the equation provided below.

$$\frac{x^5}{x^2 - 6x} + \frac{5}{x} = ?$$

A)
$$\frac{4 + x^6}{x^2 - 3x}$$

B)
$$\frac{4x^2 - 16x}{x^7}$$

C)
$$\frac{x^5 + 5x + 30}{x^2 - 6x}$$

D)
$$\frac{x^5 + 5x - 30}{x^2 + 6x}$$

E)
$$\frac{x^5 + 5x - 30}{x^2 - 6x}$$

28) One-half inch on a map represents *M* miles. Which of the following equations represents *M* + 5 miles on the map?

A) $\frac{M+5}{2M}$

B) $\frac{0.5M+2.5}{M}$

C) $\frac{2M+5}{M}$

D) $\frac{M+5}{2}$

E) $\frac{1}{2}M + 5$

29) Look at the table below and then answer the question that follows.

	Live in the City	Live in the Country
Factory workers	35	23
Office workers	42	20

The table shows data on the distribution of residents of a particular county, according to their types of work and the locations of their residences. If one of the residents from this sample is selected at random, what is the probability that the resident will be an office worker who lives in the country?

A) $^1/_7$
B) $^1/_6$
C) $^1/_3$
D) $^{23}/_{120}$
E) $^{42}/_{120}$

30) Consider a data set of integers that are greater than 50 and less than 100? What is the range of numbers that are multiples of 5 in this data set?

A) 100
B) 50
C) 45
D) 40
E) 35

31) In a particular school, there are children with black hair, blonde hair, and brown hair. If one child is selected at random, the probability that the child has blonde hair is $^1/_6$ and the probability that the child has black hair is $^3/_4$. What is the probability that the child has brown hair?

A) $^1/_{12}$
B) $^2/_3$
C) $^1/_3$
D) $^1/_6$
E) $^{11}/_{12}$

32) The number of visitors a museum had on Tuesday (T) was twice as much as the number of visitors it had on Monday (M). The number of visitors it had on Wednesday (W) was 20% greater than that on Tuesday. Which equation can be used to calculate the total number of visitors to the museum for the three days?

A) W + .20W + 2T + M
B) 2M + T + W
C) M + 1.2T + W
D) M + 2T + W
E) 5.4M

33) A construction company is building new homes on a housing development. It has an agreement with the municipality that H number of houses must be built every 30 days. If H number of houses are not built during the 30-day period, the company has to pay a penalty to the municipality of P dollars per house. The penalty is paid per house for the number of houses that fall short of the 30-day target. If A represents the actual number of houses built during the 30-day period, which equation below can be used to calculate the penalty for the 30-day period?

A) $(H - P) \times 30$
B) $(H - A) \times P$
C) $(A - H) \times 30$
D) $(A - H) \times P$
E) $(H - A) \times 30$

34) The journey from Andersonville to Blairstown on the Regional Railway is always exactly the same duration. The journey times are shown in the table below.

Regional Railway Train Service from Andersonville to Blairstown	
Departure Time (Andersonville)	Arrival Time (Blairstown)
9:50 am	10:36 am
11:15 am	12:01 pm
12:30 pm	
2:15 pm	3:01 pm
	5:51 pm

What is the missing arrival time in Blairstown in the chart above?
A) 1:16 pm
B) 2:05 pm
C) 2:16 pm
D) 5:05 pm
E) 5:50 pm

35) Read the information below and then answer the question that follows.

- Tom and Mary are planning a cross-country trip.
- They plan to drive 300 miles each day for seven days.
- Their car can travel 25 miles on one gallon of gasoline.
- **Problem:** How much money in total will they need to pay for gasoline during their trip?

What piece of information is needed in order to answer the problem?
A) The amount of gasoline that the tank of the car can hold.
B) The total amount of miles that they will drive that week.
C) The price per gallon of gasoline.
D) The day of the week that their journey will begin.
E) No further information is required.

36) Use the diagram below to answer the question that follows.

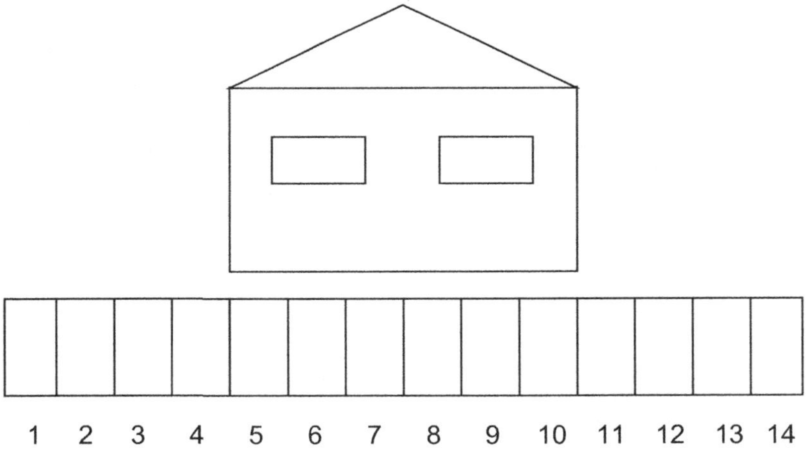

If each rectangle in the ruler below the picture of the house is one unit and the actual length of the house is 36 feet, then what is the scale of the diagram of the house?
A) 1 unit = 6 feet
B) 1 unit = 7.2 feet
C) 1 unit = 9 feet
D) 1 unit = 12 feet
E) 1 unit = 24 feet

37) A recipe of the ingredients needed to make 4 brownies is provided below.

Brownie recipe
¼ cup of flour
½ cup of sugar
¼ cup of butter
3 tablespoons of cocoa powder
¼ teaspoon of baking powder
½ teaspoon of vanilla extract

How much cocoa powder and baking powder together is needed to make 12 brownies? (1 tablespoon = 3 teaspoons)
A) 9¼ teaspoons
B) 9¾ teaspoons
C) 27¼ teaspoons
D) 27½ teaspoons
E) 27¾ teaspoons

38) Consider two concentric circles with radii of $R_1 = 1$ and $R_2 = 2.4$ as shown in the illustration below. Line L forms the diameter of the circles. What is the area of the lined part of the illustration?

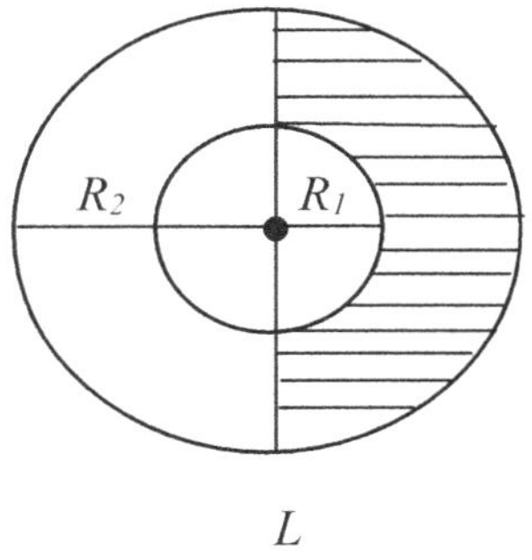

A) 0.7π
B) 1.4π
C) 2π
D) 2.38π
E) 2.8π

39) The illustration that follows represents a circular pond. The line from point X to point Y represents a bridge that passes over the exact center of the pond, and point X and Y lie on the circumference of the circle. The circumference of the pond is 176 yards. Approximately how long is the path in feet?

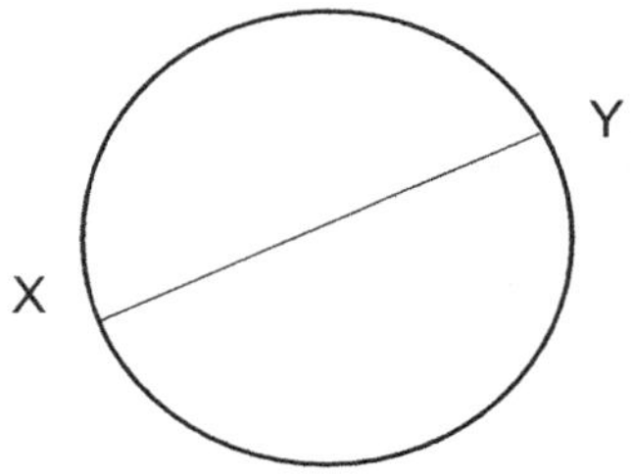

A) 28
B) 56
C) 58
D) 84
E) 168

40) The chart below shows data on the number of vehicles involved in accidents in Cedar Valley.

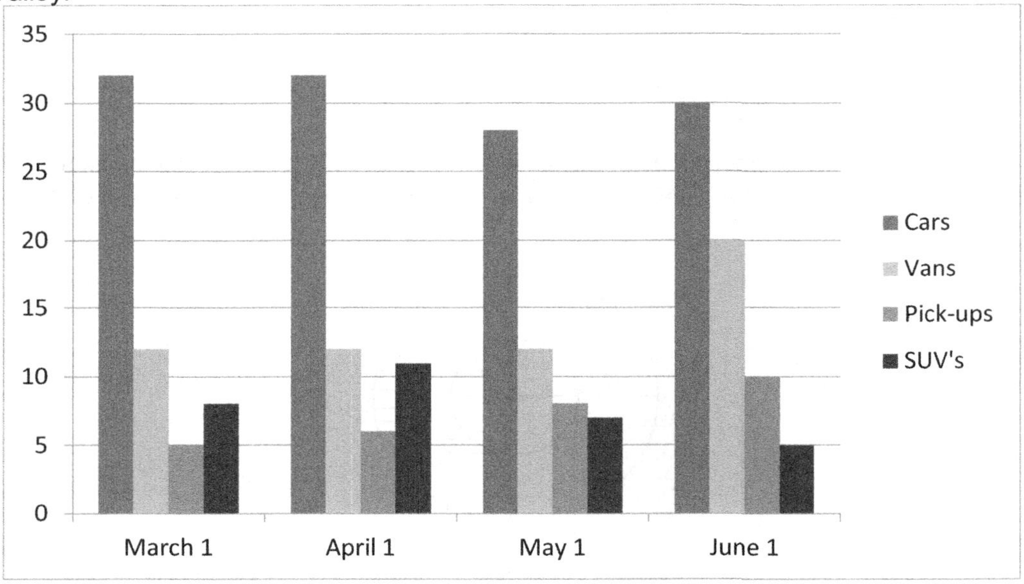

How many vans and SUV's were involved in accidents in Cedar Valley on March 1?
A) 15
B) 17
C) 20
D) 37
E) 39

41) Which of the following statements about parallelograms is true? You may select more than one response.

__________ (1) A parallelogram has no right angles.

__________ (2) A parallelogram has opposite angles which are congruent.

__________ (3) A parallelogram has two pairs of parallel sides.

__________ (4) The opposite sides of a parallelogram are unequal in measure.

__________ (5) A rectangle is not a parallelogram.

42) Which of the following statements best describes supplementary angles?
A) Supplementary angles must add up to 90 degrees.
B) Supplementary angles must add up to 180 degrees.
C) Supplementary angles must add up to 360 degrees.
D) Supplementary angles must be congruent angles.
E) Supplementary angles must be opposite angles.

43) The area of a rectangle is 168 square units. This rectangle contains smaller rectangles that measure 2 square units each. How many of these small rectangles are needed to make up the entire rectangle?
A) 13
B) 28
C) 42
D) 84
E) 168

44) Acme Packaging uses string to secure their packages prior to shipment. The string is tied around the entire length and entire width of the package, as shown in the following illustration:

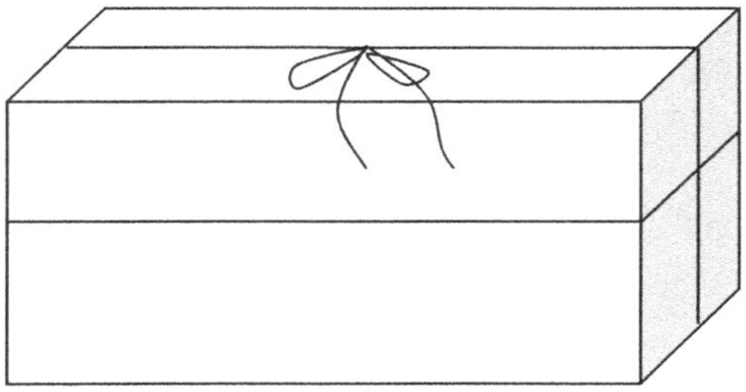

The box is ten inches in height, ten inches in depth, and twenty inches in length. An additional fifteen inches of string is needed to tie a bow on the top of the package. How much string is needed in total in order to tie up the entire package, including making the bow on the top?
A) 40
B) 80
C) 100
D) 120
E) 135

45) The triangle in the illustration below is an equilateral triangle. What is the measurement in degrees of angle a?

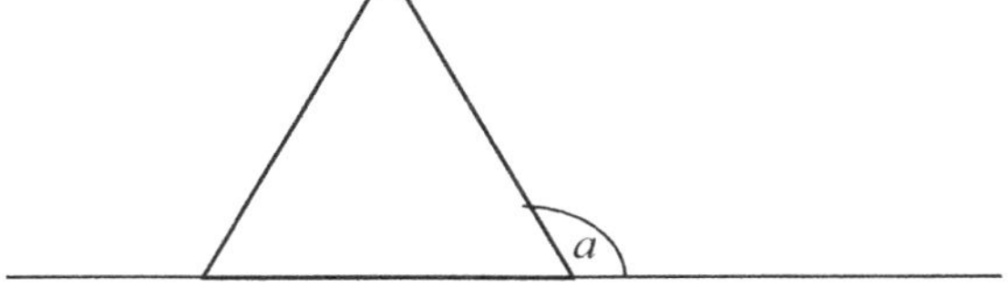

A) 40
B) 45
C) 60
D) 120
E) 130

46) 110 students took a math test. The mean score for the 60 female students was 95, while the mean score for the 50 male students was 90. Which figure below best approximates the mean test score for all 110 students in the class?
A) 55
B) 90
C) 92.5
D) 92.73
E) 95

47) Carmen wanted to find the average of the five tests she has taken this semester. However, she erroneously divided the total points from the five tests by 4, which gave her a result of 90. What is the correct average of her five tests?
A) 72
B) 85
C) 86
D) 95
E) 112.5

48) Return on investment (ROI) percentages are provided for seven companies. The ROI will be negative if the company operated at a loss, but the ROI will be a positive value if the company operated at a profit. The ROI's for the seven companies were: –2%, 5%, 7.5%, 14%, 17%, 1.3%, –3%. Which figure below best approximates the mean ROI for the seven companies?
A) 2%
B) 5.7%
C) 6.25%
D) 7.5%
E) 20%

49) A group of families had the following household incomes on their tax returns: $65000, $52000, $125000, $89000, $36000, $84000, $31000, $135000, $74000, and $87000. What is the range?
A) 74000
B) 77800
C) 79000
D) 84000
E) 104000

50) In an athletic competition, the maximum possible number of points was 25 points per participant. The scores for 15 different participants are displayed in the graph below. What was the median score for the 15 participants?

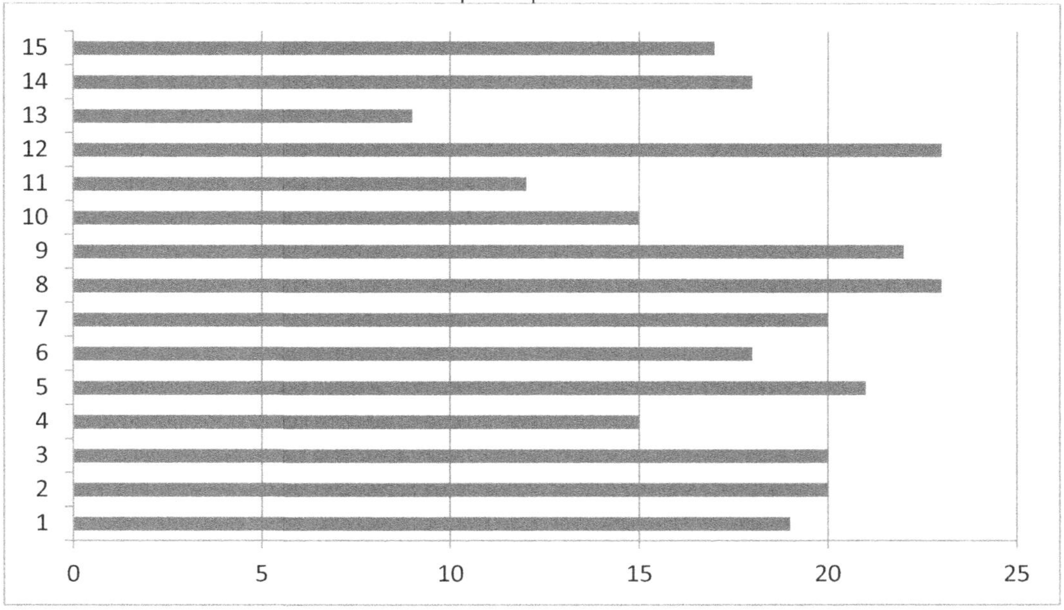

A) 15
B) 17
C) 18
D) 19
E) 23

51) A jar contains 4 red marbles, 6 green marbles, and 10 white marbles. If a marble is drawn from the jar at random, what is the probability that this marble is white?

A) $^1/_2$
B) $^1/_5$
C) $^1/_{10}$
D) $^3/_{10}$
E) $^9/_{20}$

52) Which of these numbers cannot be a probability? (There is more than one answer.)

A) –0.02
B) 0
C) 1.002
D) 1
E) $^1/_4$

53) An electricity company measures the energy consumption for each home in kilowatt hours (KWH). During July, the homes in one street had the levels of consumption in KWH in the chart show below. What is the mode of the level of energy consumption for this neighborhood for July?

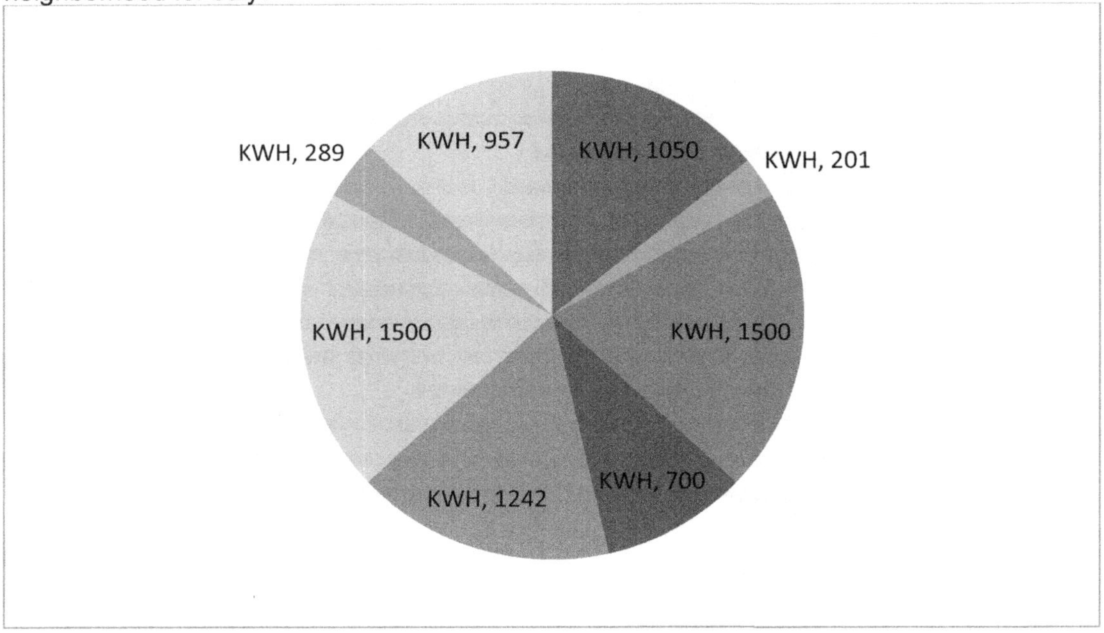

A) 700
B) 957
C) 828.5
D) 929.875
E) 1500

54) A die is rolled and a coin is tossed. What is the probability that the die shows an even number and the coin shows tails?

A) $^1/_2$
B) $^1/_4$
C) $^1/_6$
D) $^1/_{12}$
E) $^1/_{24}$

55) The blood types of 100 donors are shown in the following chart. If a donation from this group of donors is selected at random, what is the probability that type AB blood will be selected?

Blood Type	Number of donors
A positive	10
A negative	12
B positive	18
B negative	20
O type	25
AB type	15

A) $^{15}/_{99}$
B) $^{14}/_{100}$
C) $^{3}/_{20}$
D) $^{1}/_{4}$
E) $^{3}/_{5}$

56) A family is planning an annual picnic in Arizona. Rain is forecast for 45 days of the year, but when rain is forecast, the prediction is correct only 90% of the time. What is the probability that it will rain on the day of the picnic? Note that it is not a leap year.
A) 2.2222%
B) 11.0959%
C) 12.3288%
D) 45%
E) 90%

Praxis Core Math Practice Set 2 – Answer Key

1) A
2) C
3) C
4) C
5) A
6) A
7) B
8) C
9) A
10) C
11) D
12) E
13) E
14) A and C
15) A
16) B
17) E
18) A
19) 140 hours
20) D
21) D
22) A
23) D
24) D
25) E
26) A
27) E
28) B
29) B
30) D
31) A
32) E
33) B
34) A
35) C

36) A

37) E

38) D

39) E

40) C

41) (2) and (3)

42) B

43) D

44) E

45) D

46) D

47) A

48) B

49) E

50) D

51) A

52) A and C

53) E

54) B

55) C

56) B

Praxis Core Math Practice Set 2 – Solutions and Explanations

1) The correct answer is A. For problems with decimals, line the figures up in a column and add zeroes to fill in the column as shown below:

0.5400
0.0540
0.0450
0.5045
0.0054

If you still struggle with decimals, you can remove the decimal points and the zeroes before the other integers in order to see the answer more clearly.

~~0.~~5400
~~0.0~~540
~~0.0~~450
~~0.~~5045
~~0.00~~54

When we have removed the zeroes in front of the other numbers, we can see that the largest number is the first one, which is 0.54.

2) The correct answer is C. Calculate the total square footage of the room first: $12 \times 16 = 192$ total square feet. Then calculate the area under each bookcase. Bookcase A: $9 \times 1.5 = 13.5$; Bookcase B: $7 \times 1.5 = 10.5$; Total bookcase area: $13.5 + 10.5 = 24$ square feet. Next find the area of the room without the bookcase area: $192 - 24 = 168$ square feet to be covered. Finally, determine the cost: 168 square feet × \$4.25 per piece = \$714 total cost

3) The correct answer is C. Each panel is 8 feet 6 inches long, and he needs 11 panels to cover the entire side of the field. So, we need to multiply 8 feet 6 inches by 11, and then simplify the result. Step 1: 8 feet × 11 = 88 feet; Step 2: 6 inches × 11 = 66 inches; Step 3: There are 12 inches in a foot, so we need to determine how many feet and inches there are in 66 inches. 66 inches ÷ 12 = 5 feet 6 inches; Step 4: Now add the two results together. 88 feet + 5 feet 6 inches = 93 feet 6 inches

4) The correct answer is C. Circle A contains these numbers: 3, 4, 5, 6, 7, 8, 9, 10, 11, 12, 13. Circle B contains these numbers: 5, 6, 7, 8, 9, 10, 11, 12, 13, 14, 15. Now look to see which numbers are included in both data sets for region C. 5, 6, 7, 8, 9, 10, 11, 12, 13 are included in both sets. So, region C contains 9 numbers.

5) The correct answer is A. This problem is like question 1 above, except here we need to find a missing value. Remember to put in zeroes and line up the decimal points when you compare the numbers.

0.0007
A. 0.0012
B. 0.0006
C. 0.0022
D. 0.0220
E. 0.0800
0.0021

Answer choice B is less than 0.0007, and answer choices C, D, and E are greater than 0.0021. Answer choice A (0.0012) is between 0.0007 and 0.0021, so it is the correct answer.

6) The correct answer is A. The equation is: F = \$500P + \$3,700. We are told that the total funds are \$40,000 so put that in the equation to solve the problem.

\$40,000 = \$500P + \$3,700

\$40,000 – \$3,700 = \$500P

\$36,300 = \$500P

\$36,300 ÷ 500 = \$500 ÷ 500P

\$36,300 ÷ 500 = 72.6

Since a fraction of a project cannot be undertaken, the greatest number of projects is 72.

7) The correct answer is B. The dark gray part at the bottom of each bar represents those students who will attend the dance. 45% of the freshman, 30% of the sophomores, 38% of the juniors, and 30% of the seniors will attend. Calculating the average, we get the overall percentage for all four grades: (45 + 30 + 38 + 30) ÷ 4 = 35.75%.

   35% is the closest answer to 35.75%, so it best approximates our result.

8) The correct answer is C. Place the values for the temperatures in ascending order: –10, –9, 1, 6, 8, 12, 13. The median is in the middle: –10, –9, 1 , **6** , 8, 12, 13

9) The correct answer is A. The problem tells us that the morning flight had 52 passengers more than the evening flight, and there were 540 passengers in total on the two flights that day. Step 1: First of all, we need to deduct the difference from the total: 540 – 52 = 488; In other words, there were 488 passengers on both flights combined, plus the 52 additional passengers on the morning flight. Step 2: Now divide this result by 2 to allocate an amount of passengers to each flight: 488 ÷ 2 = 244 passengers on the evening flight. Had the question asked you for the number of passengers on the morning flight, you would have had to add back the number of additional passengers to find the total amount of passengers for the morning flight: 244 + 52 = 296 passengers on the morning flight

10) The correct answer is C. Divide and then round up: 82 people in total ÷ 15 people served per container = 5.467 containers. We need to round up to 6 since we cannot purchase a fractional part of a container.

11) The correct answer is D. The question is asking us about a time duration of 6 minutes, so we need to calculate the amount of seconds in 6 minutes: 6 minutes × 60 seconds per minute = 360 seconds in total. Then divide the total time by the amount of time needed to make one journey: 360 seconds ÷ 45 seconds per journey = 8 journeys. Finally, multiply the number of journeys by the amount of inches per journey in order to get the total inches: 10.5 inches for 1 journey × 8 journeys = 84 inches in total

12) The correct answer is E. First of all, we need to find a common denominator for the fractions in the inequality, as well as for the fractions in the answer choices. In order to complete the problem quickly, you should not try to find the lowest common denominator, but just find any common denominator. We can do this by expressing all of the numbers with a denominator of 90, since 9 is the largest denominator in the equation and 10 is the largest denominator in the answer choices.

    $^{2}/_{3} \times {}^{30}/_{30} = {}^{60}/_{90}$

    $^{7}/_{9} \times {}^{10}/_{10} = {}^{70}/_{90}$

    Then, express the original equation in terms of the common denominator:

    $^{60}/_{90} < ? < {}^{70}/_{90}$

    Next, convert the answer choices to the common denominator.

    A. $^{1}/_{3} \times {}^{30}/_{30} = {}^{30}/_{90}$

    B. $^{1}/_{5} \times {}^{18}/_{18} = {}^{18}/_{90}$

C. $^{2}/_{6} \times ^{15}/_{15} = ^{30}/_{90}$

D. $^{1}/_{2} \times ^{45}/_{45} = ^{45}/_{90}$

E. $^{7}/_{10} \times ^{9}/_{9} = ^{63}/_{90}$

Finally, compare the results to find the answer. By comparing the numerators (the top numbers of the fractions), we can see that $^{63}/_{90}$ lies between $^{60}/_{90}$ and $^{70}/_{90}$. So, E is the correct answer because $^{60}/_{90} < ^{63}/_{90} < ^{70}/_{90}$.

13) The correct answer is E. In ΔXYZ the long side of the triangle is 6 and ΔUWX the long side is 12. So, the ratio of the triangles is 6:12, which we simplify to 1:2 or $^{1}/_{2}$. The ratios for the two other corresponding sides are the same because the triangles are similar. So, the ratio of the other sides must also be 1:2 or $^{1}/_{2}$.

14) The correct answers are A and C. A number with seven to nine figures is in the millions. Here our number ends in zeros, so we can clearly see that a number with six zeros in the last six places of the number is in the millions. A million is equal to a thousand thousands. So, the correct answers are 120 million and 120 thousand thousands.

15) The correct answer is A. The total amount of the budget is $65,000. The up-front cost is $7,500, so we can determine the remaining amount of available funds by deducting the up-front cost from the total: $65,000 – $7,500. We have to divide the available amount by the number of employees (E) to get the maximum cost per employee: ($65,000 – $7,500) ÷ E

16) The correct answer is B. First of all, add up the number of faces on the chart: 4 + 3 + 2 + 3 = 12 faces. Each face represents 10 customers, so multiply to get the total number of customers: 12 × 10 = 120 customers in total for all four regions. The salespeople received $540 in total, so we need to divide this by the number of customers: $540 ÷ 120 customers = $4.50 per customer

17) The correct answer is E. We cannot determine the information because we would need to know exactly how much each individual customer spent in order to determine whether or not the particular customer received a free gift. We cannot simply divide and use the average.

18) The correct answer is A. The FOIL method is used on polynomials, which are equations that look like this: $(a + b)(c + d)$

You multiply the variables or terms in the parentheses in this order:

First **I**nside **O**utside **L**ast

We can use the FOIL method on our example equation as follows:

$(a + b)(c + d) =$

$(a \times c) + (a \times d) + (b \times c) + (b \times d) =$

$ac + ad + bc + bd$

You should use the FOIL method in this problem. Be very careful with the negative numbers when doing the multiplication.

$2(x + 2)(x - 3) =$

$2[(x \times x) + (x \times -3) + (2 \times x) + (2 \times -3)] =$

$2(x^2 + -3x + 2x + -6) =$

$2(x^2 - 3x + 2x - 6) =$

$2(x^2 - x - 6)$

Then multiply each term by the 2 at the front of the parentheses.

$2(x^2 - x - 6) =$

$2x^2 - 2x - 12$

19) The correct answer is 140 hours. The plumber is going to earn $4,000 for the month. She charges a set fee of $100 per job, and she will do 5 jobs, so we can calculate the total set fees first: $100 set fee per job × 5 jobs = $500 total set fees. Then deduct the set fees from the total for the month in order to determine the total for the hourly pay: $4,000 – $500 = $3,500. She earns $25 per hour, so divide the hourly rate into the total hourly pay in order to determine the number of hours she will work: $3,500 total hourly pay ÷ $25 per hour = 140 hours to work

20) The correct answer is D.

Substitute values for x to solve.

For x = –2

$x^2 + 6x + 8 = 0$

$-2^2 + (6 \times -2) + 8 = 0$

$4 - 12 + 8 = 0$

The above statement is correct, so x = –2 is one possible value.

For x = –4

$x^2 + 6x + 8 = 0$

$-4^2 + (6 \times -4) + 8 = 0$

$16 - 24 + 8 = 0$

The above statement is also correct, so x = –4 is the other possible value.

21) The correct answer is D. For questions on ratios, your first step is usually going to be to add the two parts of the ratio together: 2 + 6 = 8. So, we know that the total will have to be divisible by 8 in order to keep the total in proportion. Answer D (96) is the only answer choice that meets this criterion since 96 ÷ 8 = 12.

22) The correct answer is A. The problem tells us that A is 3 times B, and B is 3 more than 6 times C. So, we need to create equations based on this information.

B is 3 more than 6 times C: B = 6C + 3

A is 3 times B: A = 3B

Since B = 6C + 3, we can substitute 6C + 3 for B in the second equation as follows:

A = 3B

A = 3(6C + 3)

A = 18C + 9

So, A is 9 more than 18 times C.

23) The correct answer is D. The most striking relationship on the graph is the line for ages 65 and over, which clearly shows a negative relationship between exercising outdoors and the number of days of rain per month. You will recall that a negative relationship exists when an increase in one variable causes a decrease in the other variable. So, we can conclude that people aged 65 and over seem less inclined to exercise outdoors when there is more rain.

24) The correct answer is D. Divide the 45 total prescriptions by the amount stated in the rate (3 prescriptions): 45 ÷ 3 = 15. Then multiply this amount by the time stated in the rate (20 minutes) to solve: 15 × 20 = 300 minutes. Finally express in hours, if possible: 300 minutes ÷ 60 minutes per hour = 5 hours

25) The correct answer is E. When looking at scatterplots, try to see if the dots are roughly grouped into any kind of pattern or line. If so, positive or negative relationships may be represented. Here, however, the dots are located at what appear to be random places in all four quadrants of the graph. So, the scatterplot suggests that there is no relationship between x and y.

26) The correct answer is A. As x decreases by 5, y increases by 5. So, if we want to determine the x coordinate for (x, 45) we need to deduct 10 from the x coordinate of (0, 35). Therefore, the coordinates are (–10, 45), and the answer is –10.

27) The correct answer is E. Find the lowest common denominator. Since x is common to both denominators, we can convert the denominator of the second fraction to the LCD by multiplying by $(x-6)$.

$$\frac{x^5}{x^2-6x}+\frac{5}{x}=$$

$$\frac{x^5}{x^2-6x}+\left(\frac{5}{x}\times\frac{x-6}{x-6}\right)=$$

$$\frac{x^5}{x^2-6x}+\frac{5x-30}{x^2-6x}=$$

$$\frac{x^5+5x-30}{x^2-6x}$$

28) The correct answer is B. The ratio of 0.5 inch for M miles can be represented mathematically as $\frac{0.5}{M}$. The ratio for M + 5 is not known, so we can represent the unknown as x: $\frac{x}{M+5}$. Finally, use cross multiplication to solve the problem:

$$\frac{0.5}{M}=\frac{x}{M+5}$$

$$0.5\times(M+5)=Mx$$

Then divide by M to isolate x and solve the problem.

$$[0.5\times(M+5)]\div M=Mx\div M$$

$$\frac{0.5M+2.5}{M}=x$$

29) The correct answer is B. First add up all of the amounts to get the total amount: 35 + 42 + 23 + 20 = 120. We divide the number of office workers who live in the country, which is 20 people, into this total to get our answer. We can find this amount in the lower right-hand corner of the chart. So, set up the fraction and then simplify: 20 ÷ 120 = $^{20}/_{120}$ = $^{2}/_{12}$ = $^{1}/_{6}$

30) The correct answer is D. The question is asking us for numbers greater than 50 and less than 100. The number 55 is the first multiple of 5 that is greater than 5, and 95 is the first multiple of 5 that is less than 100. The range is the difference between two numbers, found by subtracting the lower number from the higher number. So, subtract to solve: 95 – 55 = 40

31) The correct answer is A. Find the lowest common denominator on the fractions. Here our common denominator is 12. For the children with blonde hair, the fraction with the common denominator is: $^1/_6 \times {}^2/_2 = {}^2/_{12}$ and for the children with black hair, it is $^3/_4 \times {}^3/_3 = {}^9/_{12}$. Then add these amounts together: $^2/_{12} + {}^9/_{12} = {}^{11}/_{12}$. The rest of the children have brown hair. The entire class is represented by the fraction $^{12}/_{12}$, so subtract to solve: $^{12}/_{12} - {}^{11}/_{12} = {}^1/_{12}$

32) The correct answer is E. Set up each part of the problem as an equation. The museum had twice as many visitors on Tuesday (T) as on Monday (M), so T = 2M. The number of visitors on Wednesday exceeded that of Tuesday by 20%, so W = 1.20 × T. Then express T in terms of M for Wednesday's visitors: W = 1.20 × T = 1.20 × 2M = 2.40M. Finally, add the amounts together for all three days: M + 2M + 2.40M = 5.4M

33) The correct answer is B. First, we need to calculate the shortage in the number of houses actually built. If H represents the number of houses that should be built and A represents the actual number of houses built, then the shortage is calculated as: $H - A$. The company has to pay P dollars per house for the shortage, so we calculate the total penalty by multiplying the shortage by the penalty per house: $(H - A) \times P$

34) The correct answer is A. Each journey lasts 46 minutes, so if the train departs at 12:30 pm, it will arrive at 1:16 pm. (12:30 + 46 minutes = 1:16)

35) The correct answer is C. In order to solve this problem, we would need to multiply the number of gallons of gasoline used each day by the cost of gasoline per gallon to calculate the cost per day. We would then add the daily amounts together to get the total cost during their entire trip. From these required facts, we are lacking the price of gasoline per gallon.

36) The correct answer is A. Count the number of units that the house spans, rather than trying to subtract units from the total of 14. If we count the number of units below the house in the drawing, we can see that the house spans 6 units. Divide this result into the actual length of the house (36 feet) to get the scale of the drawing. 36 feet ÷ 6 units = 6 feet represented by each unit

37) The correct answer is E. 3 tablespoons of cocoa powder and ¼ teaspoon of baking powder are needed for the original recipe to make 4 brownies. There are 3 teaspoons in a tablespoon, so calculate the total teaspoons needed for the original recipe first: 3 tablespoons × 3 = 9 teaspoons cocoa powder + ¼ teaspoon baking powder = 9¼ teaspoons in total. We are now making 12 brownies, so we need to multiply all of the ingredients by 3: 9¼ × 3 = 27¾ teaspoons

38) The correct answer is D. The circles are concentric, meaning that they share the same center. The formula for the area of a circle is: $\pi \times R^2$. First, we need to calculate the area of the larger circle: $\pi \times 2.4^2 = 5.76\pi$. Then calculate the area of the smaller inner circle: $\pi \times 1^2 = \pi$. We need to find the difference between half of each circle, so divide the area of each circle by 2 and then subtract:

$$(5.76\pi \div 2) - (\pi \div 2) = \frac{5.76\pi}{2} - \frac{\pi}{2} = \frac{4.76\pi}{2} = 2.38\pi$$

39) The correct answer is E. For this question, you need the formula for the circumference of a circle: $\pi \times$ diameter = circumference. If you do not see π in any of the answer choices, use 3.14 for π to solve. We are given the circumference, so we need to divide the circumference by π or 3.14 to get the diameter in yards: 176 ÷ 3.14 = 56.05, which we round down to 56 yards. There are three feet in a yard, so multiply to get the final answer: 56 × 3 = 168

40) The correct answer is C. Vans were involved in 12 accidents on March 1, and SUV's were involved in 8 accidents on the same date. So, vans and SUV's had 20 accidents in total on this date.

41) The correct answers are (2) and (3). A parallelogram is a four-sided figure that has two pairs of parallel sides. The opposite or facing sides of a parallelogram are of equal length and the opposite angles of a parallelogram are of equal measure. You will recall that congruent is another word for equal in measure. So, answers 2 and 3 are correct. A rectangle is a parallelogram with four angles of equal size (all of which are right angles), while a square is a parallelogram with four sides of equal length and four right angles.

42) The correct answer is B. Two angles are supplementary if they add up to 180 degrees.

43) The correct answer is D. A rectangle consisting of 2 square units will look like the following illustration: 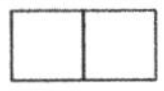

So, we divide the total number of squares in the larger rectangle by 2: 168 ÷ 2 = 84

44) The correct answer is E. The string that goes around the front, back, and sides of the package is calculated as follows: 20 + 10 + 20 + 10 = 60. The string that goes around the top, bottom, and sides of the package is calculated in the same way since the top and bottom are equal in length to the front and back: 20 + 10 + 20 + 10 = 60. So, 120 inches of string is needed so far. Then, we need 15 extra inches for the bow: 120 + 15 = 135

45) The correct answer is D. An equilateral triangle has three equal sides and three equal angles. Since all 3 angles in any triangle need to add up to 180 degrees, each angle of an equilateral triangle is 60 degrees (180 ÷ 3 = 60). Angles that lie along the same side of a straight line must add up to 180. So, we calculate angle a as follows: 180 – 60 = 120

46) The correct answer is D. You need to find the total points for all the females and the total points for all the males: Females: 60 × 95 = 5700; Males: 50 × 90 = 4500. Then add these two amounts together and divide by the total number of students in the class to get your solution: (5700 + 4500) ÷ 110 = 10,200 ÷ 110 = 92.73 average for all 110 students

47) The correct answer is A. First you need to find the total points that the student earned. You do this by taking Carmen's erroneous average times 4: 4 × 90 = 360. Then divide the total points earned by the correct number of tests in order to get the correct average: 360 ÷ 5 = 72

48) The correct answer is B. The mean is the arithmetic average. First, add up all of the items: –2% + 5% + 7.5% + 14% + 17% + 1.3% + –3% = 39.8%. Then divide by 7 since there are 7 companies in the set: 39.8% ÷ 7 = 5.68% ≈ 5.7%

49) The correct answer is E. The range is the highest number minus the lowest number. Our data set is: $65000, $52000, $125000, $89000, $36000, $84000, $31000, $135000, $74000, and $87000. So, the range is: $135000 – $31000 = $104000

50) The correct answer is D. The median is the number that is halfway through the set. Reading the graph from the bottom to top, our data set is: 19, 20, 20, 15, 21, 18, 20, 23, 22, 15, 12, 23, 9, 18, 17. First, put the numbers in ascending order: 9, 12, 15, 15, 17, 18, 18, 19, 20, 20, 20, 21, 22, 23, 23. We have 15 numbers, so the 8th number in the set is halfway and is therefore the median:

9, 12, 15, 15, 17, 18, 18, **19**, 20, 20, 20, 21, 22, 23, 23

51) The correct answer is A. Your first step is to calculate the total amount of items in the data set: 4 red marbles + 6 green marbles + 10 white marbles = 20 marbles in total. The probability is expressed with the subset in the numerator and the total remaining data set in the denominator. So, the chance of drawing a white marble is: $^{10}/_{20} = {}^{1}/_{2}$

52) The correct answers are A and C. Probability will be 1 for a 100% probability, 0 for something that has no change of occurring, or a positive number less than 1 for all other probabilities. Probability can be expressed as a decimal or a fraction. Probability therefore cannot be a negative number or a number greater than 1.

53) The correct answer is E. The mode is the number in the set that occurs most frequently. Our data set is: 1050, 201, 1500, 700, 1242, 1500, 289, 957. The number 1500 is the only number that occurs more than once, so it is the mode.

54) The correct answer is B. The data set can be expressed as follows:

(1,H),(2,H),(3,H),(4,H),(5,H),(6,H), (1,T),(2,T),(3,T),(4,T),(5,T),(6,T)

Counting the items in the above set, we can see that there are 12 items in total.

The desired outcome is that the die shows an even number and the coin shows tails.

The possible outcomes are: {(2,T),(4,T),(6,T)}

So, the probability is: $^3/_{12} = {^1/_4}$

55) The correct answer is C. The total for the data set is: 10 + 12 + 18 + 20 + 25 + 15 = 100. There are 15 donors with type AB blood, so the probability is $^{15}/_{100} = {^3/_{20}}$

56) The correct answer is B. The event is defined as the chance of rain. In terms of probabilities, we know that there are 365 days in non-leap years, so this goes in the denominator. The chance of rain goes in the numerator: $^{45}/_{365} = 12.3288\%$. However, the forecast is correct only 90% of the time: $12.3288\% \times 90\% = 11.0959\%$

Praxis Core Math Practice Set 3

1) Pilar does the following calculation:

10 kilometers × 1000 meters/1 kilometer × 100 centimeters/1 meter

What conversion is she doing?

A) Meters to kilometers
B) Centimeters to meters
C) Kilometers to meters
D) Kilometers to centimeters
E) Centimeters to kilometers

2) Which of the following shows the numbers ordered from least to greatest?

A) 0.2135
0.3152
0.0253
0.0012

B) 0.3152
0.2135
0.0253
0.0012

C) 0.0253
0.0012
0.3152
0.2135

D) 0.0012
0.0253
0.2135
0.3152

E) 0.3152
0.2135
0.0012
0.0253

3) If $\frac{x}{24}$ is between 8 and 9, which of the following could be the value of x?

A) 190
B) 191
C) 200
D) 217
E) 220

4) The ratio of bags of apples to bags of oranges in a particular grocery store is 2 to 3. If there are 44 bags of apples in the store, how many bags of oranges are there?

A) 33
B) 48
C) 55
D) 63
E) 66

5) Ali uses a jar of coffee every 7 days. Approximately how many jars of coffee will he need to last the entire year?
A) 48
B) 50
C) 52
D) 54
E) 55

6) At the beginning of class, $^1/_5$ of the students leave to go to singing lessons. Then $^1/_4$ of the remaining students leave to go to the principal's office. If 18 students are then left in the class, how many students were there at the beginning of class?
A) 90
B) 45
C) 30
D) 25
E) 24

7) In the last step of doing a calculation, Wei Li added 92 instead of subtracting 92. What shortcut can Wei Li perform in order to get the correct calculation?
A) Subtract 46 from his erroneous result.
B) Add 92 to his erroneous result.
C) Subtract 92 from his erroneous result.
D) Add 184 to his erroneous result.
E) Subtract 184 from his erroneous result.

8) Chantelle took a test that had four parts. The total number of questions on each part is given in the table below, as is the number of questions that Chantelle answered correctly.

Part	Total Number of Questions	Number of Questions Answered Correctly
1	15	12
2	25	20
3	35	32
4	45	32

Which fraction below best represents the relationship of Chantelle's incorrect answers on Part 1 to the total points on Part 1?
A) 1/5
B) 1/3
C) 4/5
D) 4/6
E) 5/6

9) A dance academy had 300 students at the beginning of January. It lost 5% of its students during the month. However, 15 new students joined the academy on the last day of the month. If this pattern continues for the next two months, how many students will there be at the academy at the end of March?
A) 285
B) 300
C) 310
D) 315
E) 320

10) In a group of children, one-half have had a tetanus shot. Of that half, only one-third suffered wounds that would have caused tetanus. In which of the following graphs does the dark gray area represent that third of the group?

A)

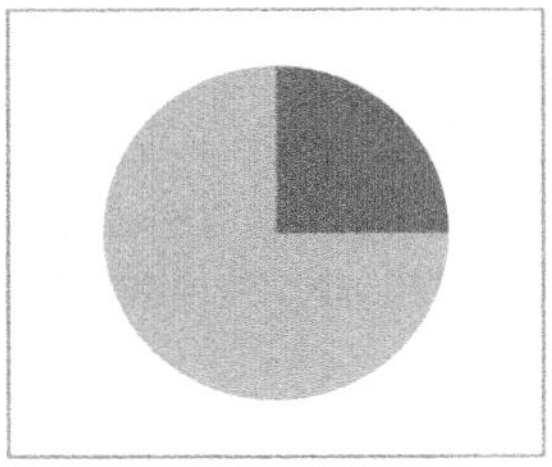

B)

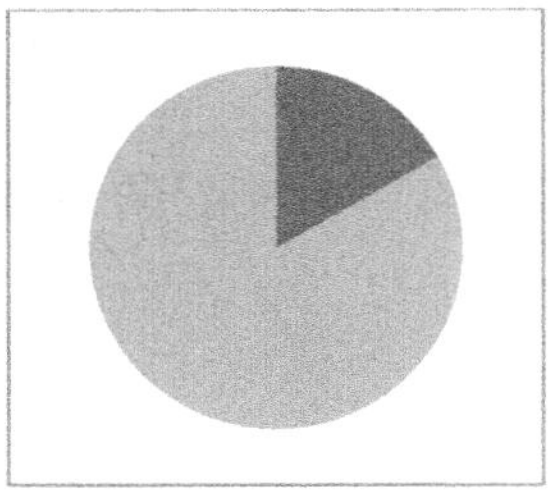

C)

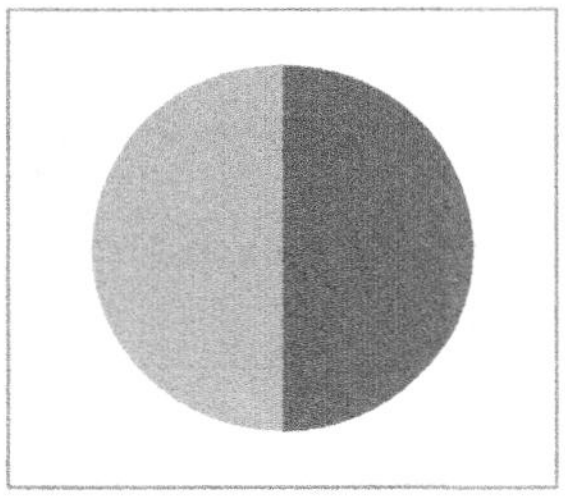

D)

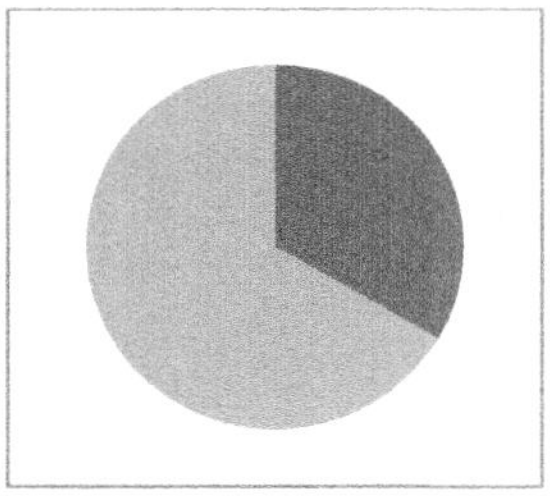

E)

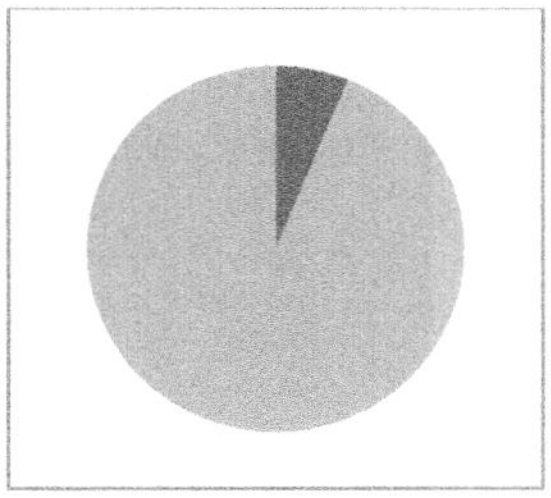

11) The residents of Hendersonville took a census. As part of the census, each resident had to indicate how many relatives they had living within a ten-mile radius of the town. The results of that particular question on the census are represented in the graph below.

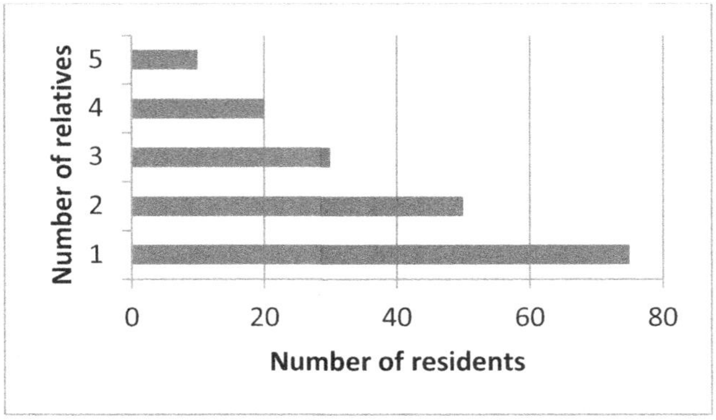

How many residents of Hendersonville had more than 3 relatives living within a ten-mile radius of the town?

A) 10
B) 20
C) 30
D) 155
E) 175

12) In the isosceles triangle RST shown below, angle T is congruent to angle R. If angle T is 49°, what is the measurement in degrees of angle S?

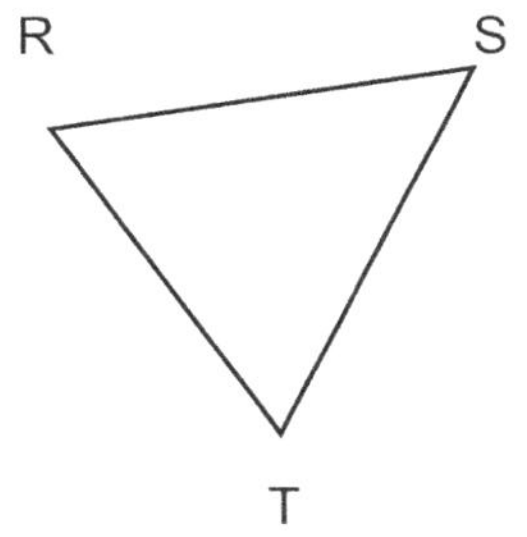

A) 49
B) 82
C) 98
D) 108
E) 131

13) A motorcycle traveled 38.4 miles in $^4/_5$ of an hour. What was the speed of the motorcycle in miles per hour?

A) 9.6
B) 30.72
C) 48
D) 52
E) 60

14) A factory that makes microchips produces 20 times as many functioning chips as defective chips. If the factory produced 11,235 chips in total last week, how many of them were defective?
A) 535
B) 561
C) 1,070
D) 10,700
E) 11,215

15) A town has recently suffered a flood. The total cost, represented by variable C, which is available to accommodate *R* number of residents in emergency housing is represented by the equation C = \$750R + \$2,550. If the town has a total of \$55,000 available for emergency housing, what is the greatest number of residents that it can house?
A) 68
B) 69
C) 70
D) 71
E) 75

16) The numbers in the following list are ordered from least to greatest:

$$\alpha,\ ^{2}/_{7},\ ^{8}/_{9},\ 1.35,\ ^{11}/_{3},\ \mu$$

Which of the following could be the value of μ? Be sure to choose all possible answers.
A) 3.5
B) $^{10}/_{4}$
C) 4.1
D) $^{1}/_{6}$
E) $^{3}/_{7}$

17) The pictograph below shows the number of traffic violations that occur every week in a certain city. The fine for speeding violations is \$50 per violation. The fine for other violations is \$20 per violation. The total collected for all three types of violations was \$6,000. What is the fine for each parking violation?

A) \$20
B) \$30
C) \$40
D) \$100
E) \$140

18) Sam is driving a truck at 70 miles per hour. He will drive through four towns on his route: Brownsville, Dunnstun, Farnam, and Georgetown. At 10:30 am, he sees this sign:

Brownsville	35 miles
Dunnstun	70 miles
Farnam	140 miles
Georgetown	210 miles

He continues driving at the same speed. What time will Sam arrive in Georgetown if he takes a 30-minute break in Farnam?
A) 1:00 pm
B) 1:30 pm
C) 2:00 pm
D) 2:30 pm
E) 2:45 pm

19) Look at the pie chart below and answer the question that follows.

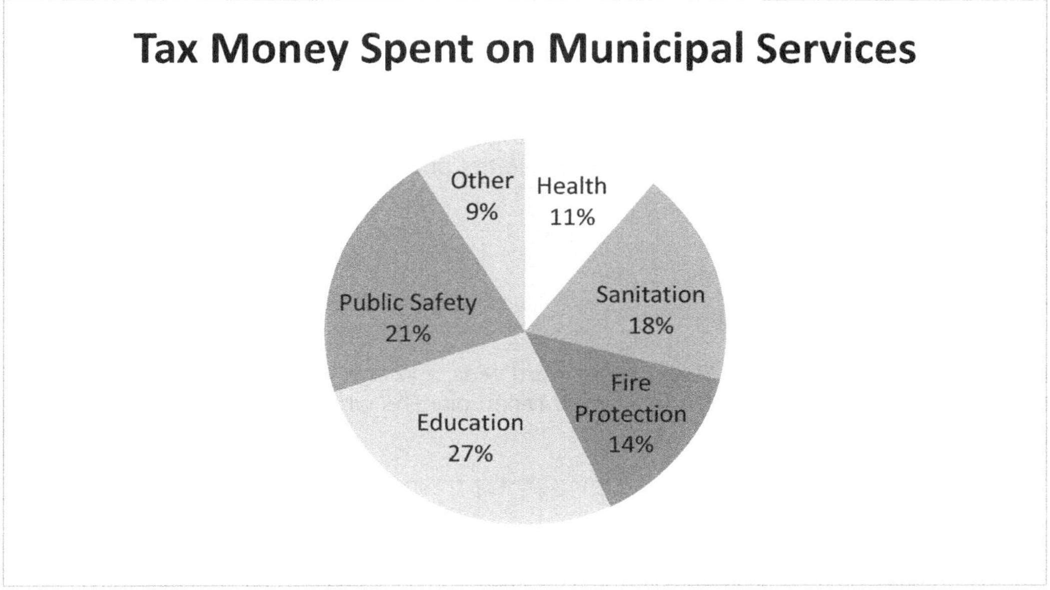

For next year, $6,537,200 in total tax money is budgeted for all municipal services. Each category is allocated the same percentage of next year's budget as the actual percentage spent for the current year. What is the budget amount for public safety?
A) $915,208
B) $1,107,813
C) $1,372,812
D) $1,765,004
E) $1,765,040

20) Look at the data below about a martial arts class and answer the question that follows.

There are 53 students at the beginning of the year.
• 15 students have black belts • 22 have brown belts • 8 have blue belts • 8 have belts of other colors

At the end of the year:
• 3 of the students with brown belts have dropped out of the class • 2 of the students with belts of other colors have also dropped out of the class • 4 new students have joined the class

Which of the following facts can be determined from the information provided?
A) The total number of students in the class.
B) The number of students in the class with brown belts.
C) The number of students in the class with blue belts.
D) The number of students in the class with black belts.
E) No facts can be determined since there is not sufficient information.

21) The price of a sofa at a local furniture store was x dollars on Wednesday this week. On Thursday, the price of the sofa was reduced by 10% of Wednesday's price. On Friday, the price of the sofa was reduced again by 15% of Thursday's price. Which of the following expressions can be used to calculate the price of the sofa on Friday?
A) $(0.25)x$
B) $(0.75)x$
C) $(0.10)(0.15)x$
D) $(0.10)(0.85)x$
E) $(0.90)(0.85)x$

22) Each square in the diagram below is one yard wide and one yard long. The gray area of the diagram represents New Town's water reservoir. The white area represents the surrounding conservation park.

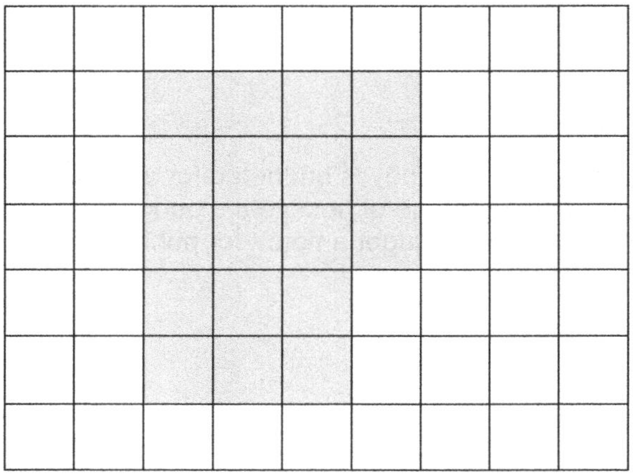

Which of the following ratios represents the area of the reservoir to the area of the surrounding conservation park?
A) 2:5
B) 9:23

C) 17:32
D) 18:44
E) 32:17

23) There are three boys in a family, named Alex, Burt, and Zander. Alex is twice as old as Burt, and Burt is one year older than three times the age of Zander. Which of the following statements best describes the relationship between the ages of the boys?
A) Alex is 4 years older than 6 times the age of Zander.
B) Alex is 2 years older than 6 times the age of Zander.
C) Alex is 4 years older than 3 times the age of Zander.
D) Alex is 2 years older than 3 times the age of Zander.
E) Alex is 1 year older than 6 times the age of Zander.

24) You need to fill a rectangular solid container with a liquid substance. The length of the rectangular solid is 12 feet, the width is 9 feet, and the volume is 1080 cubic feet. What is the height of the rectangular solid?
A) 10 feet
B) 12 feet
C) 90 feet
D) 100 feet
E) 120 feet

25) If $\frac{1}{5}x + 3 = 5$, then x = ?

A) $\frac{8}{5}$

B) $-\frac{8}{5}$

C) 8

D) 10

E) –10

26) The graph of a linear equation is shown below. Which one of the tables of values best represents the points on the graph?

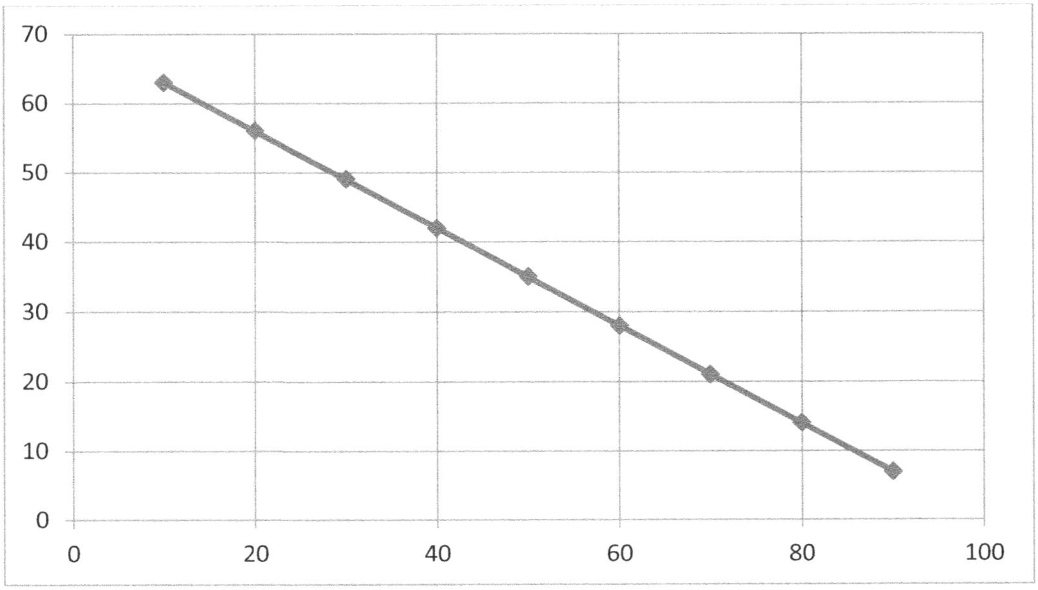

A)

x	y
5	65
10	64
15	63
20	62

B)

x	y
5	68
15	60
25	52
35	54

C)

x	y
10	63
20	56
30	49
40	42

D)

x	y
10	68
20	60
30	52
40	44

E)

x	y
30	42
40	35
50	28
60	21

27) An athlete ran 10 miles in 1.5 hours. The graph below shows the miles the athlete ran every 10 minutes. According to the graph, how many miles did the athlete run in the first 30 minutes?

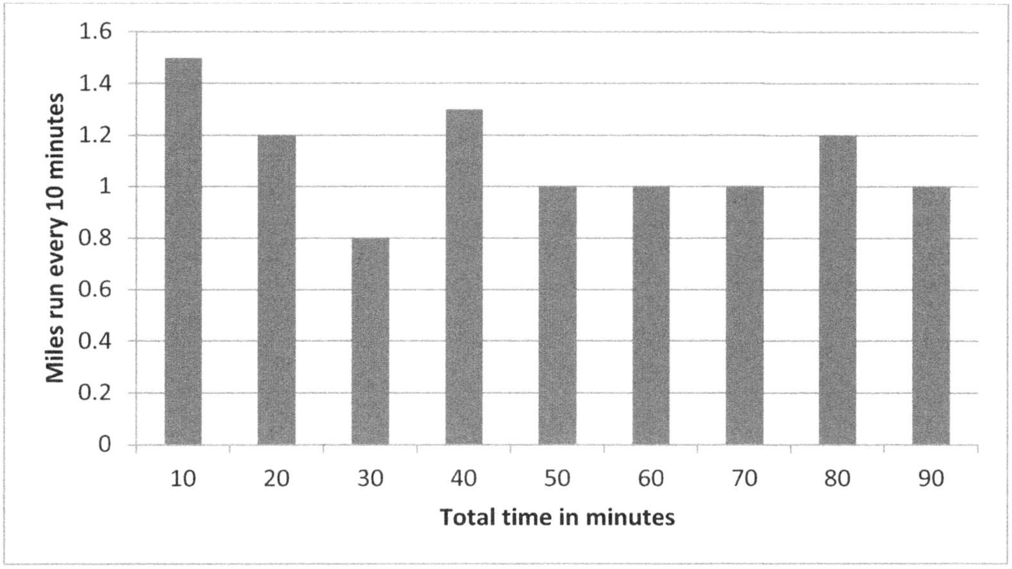

A) 0.8 miles
B) 2.0 miles
C) 3.0 miles
D) 3.4 miles
E) 3.5 miles

28) Which of the following steps will solve the equation for x: $18 = 3(x + 5)$
A) Subtract 5 from each side of the equation, and then divide both sides by 3.
B) Subtract 18 from each side of the equation, and then divide both sides by 5.
C) Multiply both x and 5 by 3 on the right side of the equation. Then subtract 15 from each side of the equation.
D) Divide each side of the equation by 3. Then subtract 5 from both sides of the equation.
E) Divide each side of the equation by 5. Then subtract 3 from both sides of the equation.

29) A packaging company secures their packages with plastic strapping prior to shipment. The box is 18 inches in height, 16 inches in depth, and 38 inches in length. For certain packages, 14 extra inches of strapping is used to make a handle on the top of the package to carry it. The strapping is wrapped around the length and width of the entire package, as shown in the following diagram:

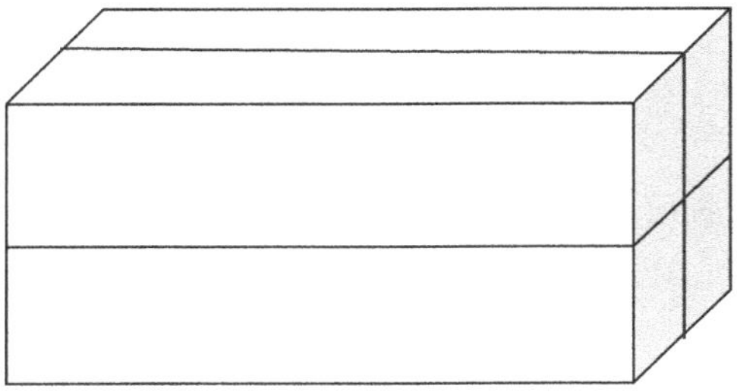

How many inches of strapping is needed to wrap 35 packages if no handles are used?
A) 5,040
B) 7,700
C) 8,190
D) 10,080
E) 10,570

30) A clothing store sells jackets and jeans at a discount during a sales period. T represents the number of jackets sold and N represents the number of jeans sold. The total amount of money the store collected for sales of jeans and jackets during the sales period was $4,000. The amount of money earned from selling jackets was one-third of that earned from selling jeans. The jeans sold for $20 a pair. How many pairs of jeans did the store sell during the sales period?
A) 15
B) 20
C) 150
D) 200
E) 3000

31) Mardetta is putting money into her checking account every week, and this weekly deposit is represented as x. After 40 weeks of putting money into her account, she took out or withdrew $180 from the account. The remaining balance on the account after the withdrawal was $1,020. Which equation below correctly expresses the balance of the account in $\$D$ before the withdrawal and the 40 weekly deposits of x were made? Note that the account earns no interest.
A) $\$D = \$1,200 + 40x$
B) $\$D = \$1,080 + 40x$
C) $\$D - 40x = \$1,200$
D) $\$D + 40x = \$1,080$
E) $\$D = \$1,200 - 40x$

32) Which of the following is equivalent to the expression $36 - 2x$ for all values of x?
A) $6 + 2(15 - x)$
B) $6(6 - 2x)$
C) $39 - (3 - 2x)$
D) $8(5 - 2x) - 4$
E) $6(6 - 4x) - 2x$

33) Carlos is going to buy a house. The total purchase price of the house is represented by variable H. He will pay D dollars immediately, and then he will make equal payments (P) each month for M months. If H = $300,000, P = $700 and M = 360, how much will Carlos pay immediately?
A) $38,000
B) $48,000
C) $58,000
D) $252,000
E) $299,300

34) Which of the following equations is equivalent to $\frac{x}{5} \div \frac{9}{y}$?
A) $\frac{xy}{45}$

B) $\frac{9x}{5y}$

C) $\frac{1}{9} \times \frac{x}{5y}$

D) $\frac{1}{5} \times \frac{9}{5y}$

E) $\frac{1}{5} \div \frac{9x}{y}$

35) The radius (R) of circle A is 5 centimeters. The radius of circle B is 3 centimeters. Which of the following statements is true? You may select more than 1 answer.

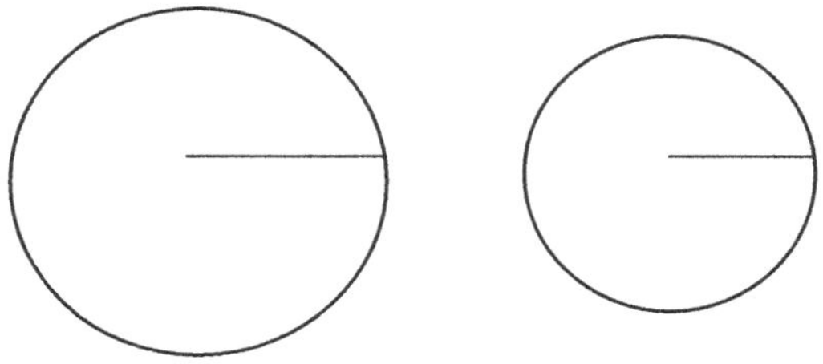

A) The difference between the areas of the circles is 2.
B) The difference between the areas of the circles is 9π.
C) The difference between the circumferences of the circles is 2.
D) The difference between the circumferences of the circles is 4π.
E) The difference between the diameters of the circles is 4.

36) Liz wants to put new vinyl flooring in her kitchen. She will buy the flooring in square pieces that measure 1 square foot each. The entire room is 8 feet by 12 feet. The cupboards are two feet deep from front to back. Flooring will not be put under the cupboards. A diagram of her kitchen is provided.

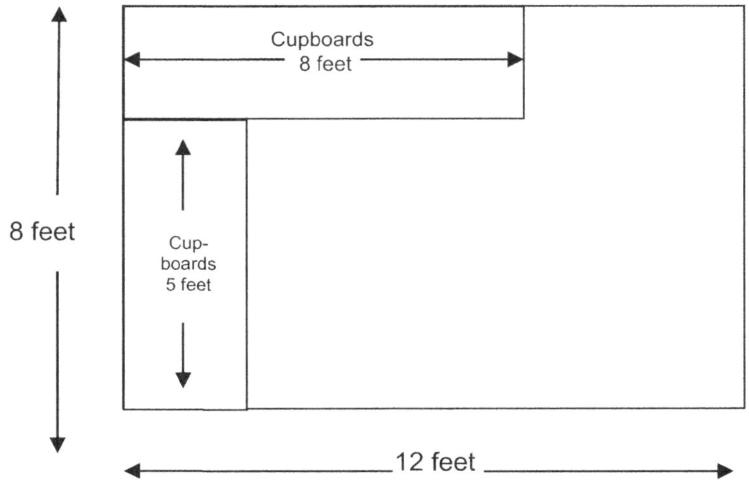

How many pieces of vinyl will Liz need to cover her floor?
A) 120
B) 96
C) 70
D) 84
E) 88

37) A large wheel (L) has a radius of 10 inches. A small wheel (S) has a radius of 6 inches. If the large wheel is going to travel 360 revolutions, how many more revolutions does the small wheel need to make to cover the same distance?
A) 120
B) 240
C) 360
D) 720
E) 120π

38) Look at the bar chart below and answer the question that follows.

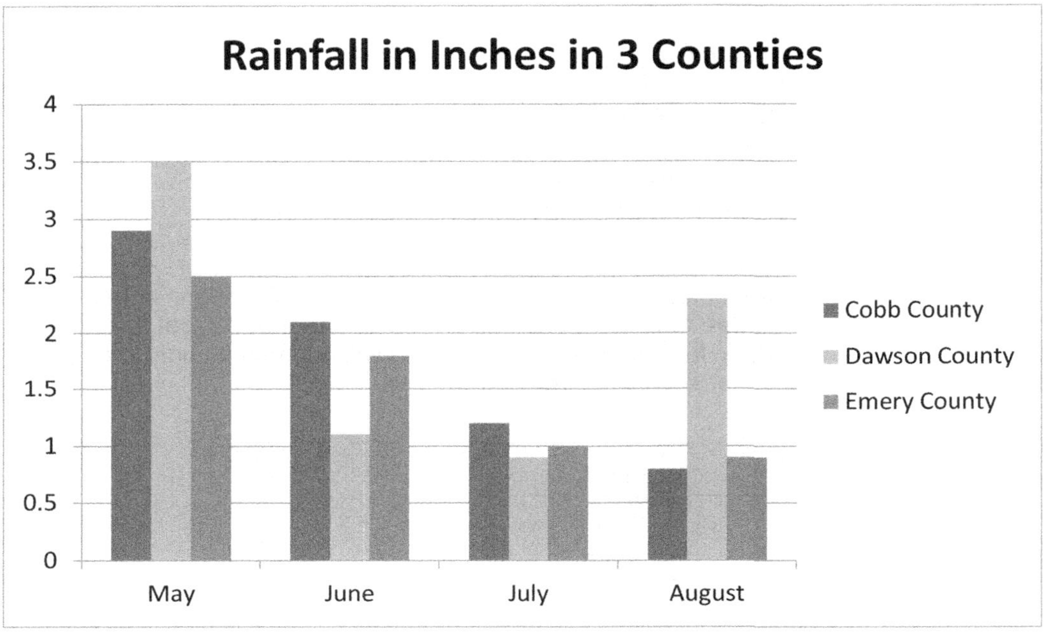

What was the approximate difference in the amount of rainfall for Dawson County and Emery County for June?
A) Dawson County had 0.6 more inches of rainfall than Emery County.
B) Emery County had 0.6 more inches of rainfall than Dawson County.
C) Dawson County had 1.1 fewer inches of rainfall than Emery County.
D) Emery County had 1.1 fewer inches of rainfall than Dawson County.
E) Emery County had 1.6 fewer inches of rainfall than Dawson County.

39) A farmer has a rectangular pen in which he keeps animals. He has decided to divide the pen into two parts. To divide the pen, he will erect a fence diagonally from the two corners, as shown in the diagram below. How long in yards is the diagonal fence?

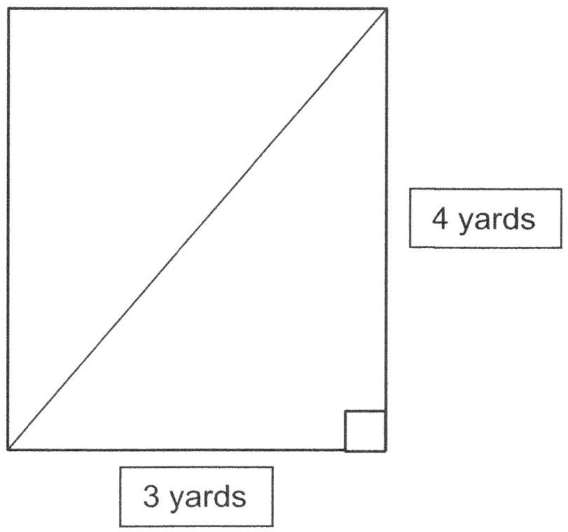

A) 4
B) 5
C) 5.5
D) 6
E) 6.5

40) The diagram below shows a figure made from a semicircle, a rectangle, and an equilateral triangle. The rectangle has a length of 18 inches and a width of 10 inches. What is the perimeter of the figure?

A) 56 inches + 5π inches
B) 56 inches + 10π inches
C) 56 inches + 12.5π inches
D) 56 inches + 25π inches
E) 208.9 inches + 12.5π inches

41) Read the problem below and then answer the question that follows.

- Paul leaves his house at 5:30 to go running.
- He runs 2 miles north through town, then continues 3 miles further north past town.
- He then runs south to his house along the same route.
- What is Paul's running pace?

What piece of information is needed in order to answer the problem?
A) The number of steps that Paul makes.
B) The time that Paul returns home.
C) The length of Paul's stride.

D) The length of the return journey.
E) No further information is required.

42) Look at the table below and answer the question that follows.

Disease or Complication	Percentage of patients with this disease that have survived and total number of patients
Cardiopulmonary and vascular	82% (602,000)
HIV/AIDS	73% (215,000)
Diabetes	89% (793,000)
Cancer and leukemia	48% (231,000)
Premature birth complications	64% (68,000)

The total number of deaths from the two least fatal diseases amounted to which figure below?

A) 82,530
B) 208,960
C) 1,186,040
D) 1,199,410
E) 1,919,410

43) The illustration below shows a pentagon. The shaded part at the top of the pentagon has a height of 6 inches.

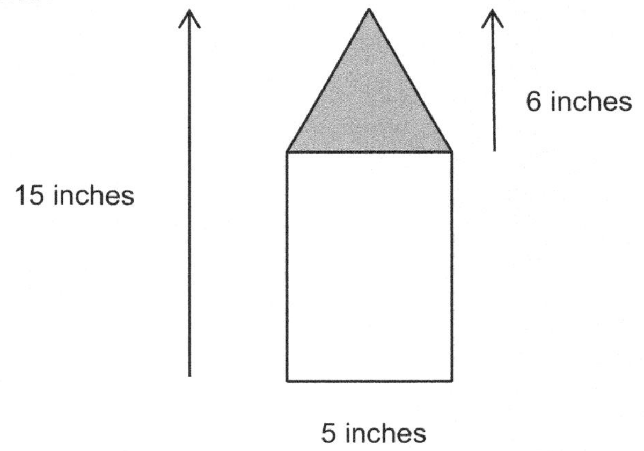

The height of the entire pentagon is 15 inches, and the base of the pentagon is 5 inches. What fraction expresses the area of the shaded part to the area of the entire pentagon? State your answer as a simplified fraction in the spaces provided.

Numerator value = ___________

Denominator value = __________

44) Look at the graph below and answer the question that follows.

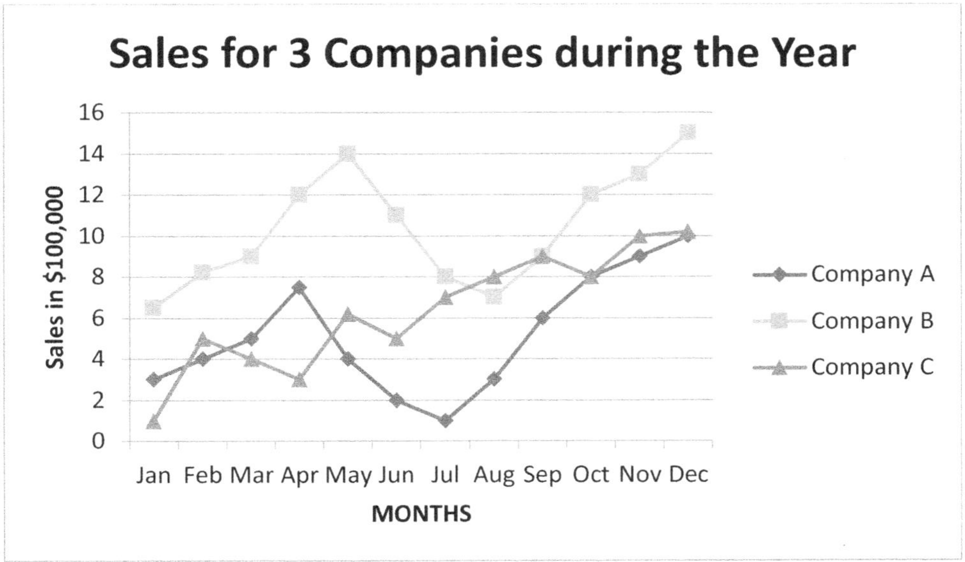

The combined total of sales for all three of the companies was greatest during which month of the year?

A) December
B) November
C) May
D) April
E) January

45) In the figure below, *x* and *y* are parallel lines, and line *z* is a transversal crossing both *x* and *y*. Which three angles are equal in measure? You may select more than one answer.

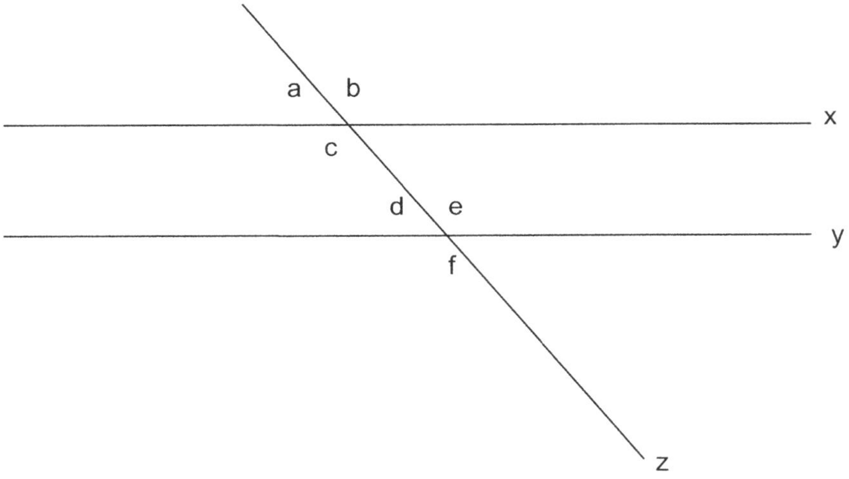

A) $\angle a$, $\angle b$, $\angle c$
B) $\angle b$, $\angle c$, $\angle f$
C) $\angle b$, $\angle e$, $\angle f$
D) $\angle a$, $\angle d$, $\angle e$
E) $\angle a$, $\angle d$, $\angle f$

46) What is the median of the numbers in the following list:
2.5, 9.4, 3.1, 1.7, 3.2, 8.2, 4.5, 6.4, 7.8
A) 3.2
B) 4.5
C) 5.2
D) 6.4
E) 7.7

47) Suki rolls a fair pair of six-sided dice. Each die has values from 1 to 6. She rolls an even number on her first roll. What is the probability that she will roll an odd number on her next roll?
A) $^1/_2$
B) $^1/_6$
C) $^2/_6$
D) $^6/_{11}$
E) $^5/_{11}$

48) The state highway department wants to find out how the residents of Buford feel about a new road being constructed around their town. Which one of the following methods will result in the most statistically valid information about the opinions of the residents at the town?
A) To question drivers of a random selection of vehicles traveling through Buford
B) To poll drivers at random as they exit the interstate highway near Buford
C) To ask car owners living in Buford to participate in a survey
D) To select a random sample of names from Buford's voting register
E) To select participants for a survey from a list of all of the citizens living in Buford

49) A student receives the following scores on his exams during the semester:
89, 65, 75, 68, 82, 74, 86
What is the mean of his scores?
A) 24
B) 74
C) 75
D) 77
E) 82

50) A clown pulls balloons out of a bag at random to blow up and give to children during a birthday party. At the start of the party, there are 10 red balloons, 7 green balloons, 6 purple balloons, 5 orange balloons, and 11 blue balloons in the bag. The clown selects the first balloon, which is blue, and gives it to the first child. If the second child gets an orange balloon, what is the probability that the third child will get a blue balloon?

A) $^{11}/_{37}$

B) $^{10}/_{37}$

C) $^{11}/_{39}$

D) $^{10}/_{39}$

E) $^1/_{11}$

51) What is the mode of the numbers in the following list: 1.6, 2.9, 4.5, 2.5, 2.6, 5.1, 5.4
A) 3.5 B) 3.1 C) 3.0 D) 2.5 E) no mode

52) Sam was adding up the following numbers. However, when he put the numbers into the calculator, he accidently swapped the digits on one of the numbers. His erroneous sum is off by 36. Which number did he enter incorrectly into his calculator?

$$\begin{array}{r} 59 \\ 82 \\ 14 \\ 35 \\ \underline{+\ 97} \end{array}$$

A) 14 B) 35 C) 59 D) 82 E) 97

53) There are 10 cars in a parking lot. Nine of the cars are 2, 3, 4, 5, 6, 7, 9, 10, and 12 years old, respectively. If the average age of the 10 cars is 6 years old, how old is the 10th car?
A) 1 year old
B) 2 years old
C) 3 years old
D) 4 years old
E) 5 years old

54) Which of the following are statistical questions? You may select more than one answer.
______ (1) What size shoe does Mrs. Shapiro wear?

______ (2) How many residents of the town oppose the tax increase?

______ (3) Will it rain tomorrow?

______ (4) How many miles can that car travel on a tank of gasoline?

______ (5) Do most college graduates in our state find jobs within one year of leaving college?

55) The range and mean of 5 numbers are 10 and 14 respectively. The five numbers are positive integers greater than 0. If the range is increased by 2, which of the following could be true of the numbers in the new set? You may select more than one answer.
A) The highest number is increased by 2.
B) The highest number is increased by 1 and the lowest number is decreased by 1.
C) The highest number is decreased by 1 and the lowest number is increased by 1.
D) The lowest number is increased by 2.
E) The mean of the numbers remains unchanged

56) A deck of cards contains 13 hearts, 13 diamonds, 13 clubs, and 13 spades. Cards are selected from the deck at random. Once selected, the cards are discarded and are not placed back into the deck. Two spades, one heart, and a club are drawn from the deck. What is the probability that the next card drawn from the deck will be a heart?
A) $^{1}/_{13}$
B) $^{1}/_{12}$
C) $^{13}/_{52}$
D) $^{13}/_{48}$
E) $^{1}/_{4}$

Praxis Core Math Practice Set 3 – Answer Key

1) D
2) D
3) C
4) E
5) C
6) C
7) E
8) A
9) B
10) B
11) C
12) B
13) C
14) A
15) B
16) C
17) C
18) C
19) C
20) A
21) E
22) A
23) B
24) A
25) D
26) C
27) E
28) D
29) B
30) C
31) E
32) A
33) B
34) A
35) D and E

36) C

37) B

38) B

39) B

40) A

41) B

42) A

43) numerator value = 1; denominator value = 4

44) A

45) B and C

46) B

47) A

48) E

49) D

50) B

51) E

52) C

53) B

54) 2 and 5

55) A and B

56) E

Praxis Core Math Practice Set 3 – Solutions and Explanations

1) The correct answer is D. Remember to look at the first item in the expression, as well as the numerator of the final item in the expression in order to understand the conversion:

   10 **kilometers** × 1000 meters/1 kilometer × 100 **centimeters**/1 meter

   In other words, she is converting kilometers to centimeters.

2) The correct answer is D. We have the following numbers in our problem:

   0.0012
   0.0253
   0.2135
   0.3152

   If you still do not feel confident with decimals, remember that you can remove the decimal point and the zeroes after the decimal but before the other integers in order to see the answer more clearly.

   12
   253
   2135
   3152

3) The correct answer is C. If $\frac{x}{24}$ is between 8 and 9, x will need to be between 192 and 216, since $\frac{192}{24}$ =192 ÷ 24 = 8 and $\frac{216}{24}$ = 216 ÷ 24 = 9. 200 is the only number from the answer choices that is greater than 192 and less than 216.

4) The correct answer is E. The ratio of bags of apples to bags of oranges is 2 to 3, so for every two bags of apples, there are three bags of oranges. First, take the total amount of bags of apples and divide by 2: 44 ÷ 2 = 22. Then multiply this by 3 to determine how many bags of oranges are in the store: 22 × 3 = 66

5) The correct answer is C. If Ali uses a jar of coffee every week, he needs 52 jars to last a year since there are 52 weeks in a year.

6) The correct answer is C. Work backwards based on the facts given. There are 18 students left at the end after one-fourth of them left for the principal's office. So, set up an equation for this:

   $18 + \frac{1}{4}T = T$

   $18 + \frac{1}{4}T - \frac{1}{4}T = T - \frac{1}{4}T$

   $18 = \frac{3}{4}T$

   $18 \times 4 = \frac{3}{4}T \times 4$

   $72 = 3T$

   $72 \div 3 = 3T \div 3$

   $24 = T$

   So, before the group of pupils left to see the principal, there were 24 students in the class. We know that one-fifth of the students left at the beginning to go to singing lessons, so we need to set up an equation for this:

   $24 + \frac{1}{5}T = T$

   $24 + \frac{1}{5}T - \frac{1}{5}T = T - \frac{1}{5}T$

$24 = {}^{4}/_{5}T$

$24 \times 5 = {}^{4}/_{5}T \times 5$

$120 = 4T$

$120 \div 4 = 4T \div 4$

$30 = T$

7) The correct answer is E. He needs to subtract the 92 that he added by mistake to get back to his starting point. Then he needs to subtract 92 again to get the correct result. So, he can subtract 92 two times or simply shortcut by subtracting 184.

8) The correct answer is A. Chantelle correctly answered 12 out of 15 questions, so she incorrectly answered 3 questions (15 – 12 = 3). This can be expressed as the fraction 3/15, which can be simplified to 1/5.

9) The correct answer is B. At the beginning of January, there are 300 students, but 5% of the students leave during the month, so we have 95% left at the end of the month: 300 × 95% = 285. Then 15 students join on the last day of the month, so we add that back in to get to the total at the end of January: 285 + 15 = 300. If this pattern continues, there will always be 300 students in the academy at the end of any month.

10) The correct answer is B. The question is asking us to calculate one third of one half. So, we multiply to get our answer: $^{1}/_{2} \times {}^{1}/_{3} = {}^{(1 \times 1)}/_{(2 \times 3)} = {}^{1}/_{6}$. This is represented by the dark gray area in graph B.

11) The correct answer is C. The question is asking us how many residents have more than 3 relatives nearby, so we need to add the bars for 4 and 5 relatives from the chart. 20 residents have 4 relatives nearby and 10 residents have 5 relatives nearby, so 30 residents (20 + 10 = 30) have more than 3 relatives nearby.

12) The correct answer is B. In an isosceles tringle, two angles will have equal measurements. These two angles are known as congruent angles. In other words, congruent means equal in measurement. So, first of all, add the two congruent angles together: 49 + 49 = 98. Remember that the sum of all of the angles in a triangle must be equal to 180 degrees. So, subtract the sum of the congruent angles from 180 to solve: 180 – 98 = 82

13) The correct answer is C. Divide by the fractional hour in order to determine the speed for an entire hour: 38.4 miles ÷ $^{4}/_{5}$ of an hour = 38.4 × $^{5}/_{4}$ = (38 × 5) ÷ 4 = 48 mph

14) The correct answer is A. The ratio of defective chips to functioning chips is 1 to 20. So, the defective chips form one group and the functioning chips form another group. Therefore, the total data set can be divided into groups of 21. Accordingly, $^{1}/_{21}$ of the chips will be defective. The factory produced 11,235 chips last week, so we calculate as follows: 11,235 × $^{1}/_{21}$ = 535

15) The correct answer is B. The total amount available is $55,000, so we can substitute this for C in the equation provided in order to calculate *R* number of residents:

C = $750*R* + $2,550

$55,000 = $750*R* + $2,550

$55,000 – $2,550 = $750*R* + $2,550 – $2,550

$55,000 – $2,550 = $750*R*

$52,450 = $750*R*

$52,450 ÷ $750 = $750*R* ÷ $750

$52,450 ÷ $750 = *R*

69.9 = *R*

It is not possible to accommodate a fractional part of one person, so we need to round down to 69 residents.

16) The correct answer is C. The value of μ must be greater than $^{11}/_3$, which is equal to 3.6667. The answer 4.1 is the only option which meets this criterion.

17) The correct answer is C. There are 2 stars for speeding, and each star equals 30 violations, so there were 60 speeding violations in total. The fine for speeding is \$50 per violation, so the total amount collected for speeding violations was: 60 speeding violations × \$50 per violation = \$3000. There are three stars for other violations, which is equal to 90 violations (3 × 30 = 90). Other violations are \$20 each, so the total for other violations is: 90 × \$20 = \$1800. Next, we need to deduct these two amounts from the total collections of \$6,000 in order to find out how much was collected for parking violations: \$6000 – \$3000 – \$1800 = \$1200 in total for parking violations. There is one star for parking violations, so there were 30 parking violations. We divide to get the answer: \$1200 income for parking violations ÷ 30 parking violations = \$40 each

18) The correct answer is C. We know that at 10:30 he is 210 miles from Georgetown. If he is traveling 70 miles per hour, he would need 3 more hours to get to Georgetown since 3 × 70 = 210. So, he would arrive at 1:30 if he had not taken a break. But we need to add in a 30-minute break, so he will arrive in Georgetown at 2:00 pm.

19) The correct answer is C. Take the total dollar amount of the budget and multiply by the 21% for public safety: \$6,537,200 × 0.21 = \$1,372,812

20) The correct answer is A. We cannot calculate the number of students in the class with belts of particular colors because we do not know the color of the belts the new students. The problem is telling us how many students there are in each group and how many of each group have left. The problem also tells us how many students in total have joined, so we can calculate the new total number of students.

21) The correct answer is E. The original price of the sofa on Wednesday was x. On Thursday, the sofa was reduced by 10%, so the price on Thursday was 90% of x or $0.90x$. On Friday, the sofa was reduced by a further 15%, so the price on Friday was 85% of the price on Thursday, so we can multiply Thursday's price by 0.85 to get our answer: $(0.90)(0.85)x$

22) The correct answer is A. First, calculate the total area represented on the diagram: 9 × 7 = 63 square yards in total. Then count the gray squares for the reservoir. We can see that there are 18 gray squares. So, we know that the gray area is 18 square yards and the white area is 45 square yards, since 63 – 18 = 45. So, the ratio is 18:45. Both of these numbers are divisible by 9, so we can simplify the ratio to 2:5 because 18 ÷ 9 = 2 and 45 ÷ 9 = 5.

23) The correct answer is B. Assign a variable for the age of each boy. Alex = A, Burt = B, and Zander = Z. Alex is twice as old as Burt, so A = 2B. Burt is one year older than three times the age of Zander, so B = 3Z + 1. Then substitute the value of B into the first equation.

A = 2B

A = 2(3Z + 1)

A = 6Z + 2

So, Alex is 2 years older than 6 times the age of Zander.

24) The correct answer is A. From the formula sheet, we know that the formula for the volume of a rectangular is: solid = length × width × height. Here, we are given the volume (the larger number of 1080), so we need to divide that by the length and then the width in order to find the height: (1080 ÷ 12) ÷ 9 = 90 ÷ 9 = 10 feet

25) The correct answer is D. Get the integers to one side of the equation first of all.

$\frac{1}{5}x + 3 = 5$

$\frac{1}{5}x + 3 - 3 = 5 - 3$

$\frac{1}{5}x = 2$

Then multiply to eliminate the fraction and solve the problem.

$\frac{1}{5}x \times 5 = 2 \times 5$

$x = 10$

26) The correct answer is C. The first point on the graph lies at x = 10, so we can eliminate answer choices A and B. The point for the y coordinate that corresponds to x = 10 is 63 not 68, so we can eliminate answer choice D. The point for the y coordinate that corresponds to x = 30 is 49 not 42, so we can also eliminate answer choice E.

27) The correct answer is E. The first three bars of the graph represent the first 30 minutes, so add these three amounts together for your answer: 1.5 + 1.2 + 0.8 = 3.5 miles

28) The correct answer is D. Divide each side of the equation by 3. Then subtract 5 from both sides of the equation as shown below.

$18 = 3(x + 5)$

$18 \div 3 = [3(x + 5)] \div 3$

$6 = x + 5$

$6 - 5 = x + 5 - 5$

$1 = x$

29) The correct answer is B. Calculate the length of strapping for the piece that goes over the front of the package: 38 + 16 + 38 + 16 = 108. Then calculate the length of strapping for the piece that goes over the top of the package: 38 + 18 + 38 + 18 = 112. Without the handle, we need 108 + 112 = 220 inches per package. 220 inches per package × 35 packages = 7,700 total inches

30) The correct answer is C. If the amount earned from selling jackets was one-third that of selling jeans, the ratio of jacket to jean sales was 1 to 3. So, we need to divide the total sales of $4,000 into $1,000 for jackets and $3,000 for jeans. We can then solve as follows: $3,000 in jeans sales ÷ $20 per pair = 150 pairs sold

31) The correct answer is E. We are being asked for the balance of the account before the withdrawal and the 40 weekly deposits. There was $1,020 in the account after the $180 withdrawal, so add the withdrawal back in to get the balance at that point in time: $1,020 + $80 = $1,200. Then subtract the amount of the deposits to get the balance at the start: $1,200 – 40$x$. So, E is the correct answer.

32) The correct answer is A. For algebraic equivalency questions like this, you can perform the operations on each of the answer choices to see which one is equivalent. Remember to be careful when performing multiplication on negative numbers inside parentheticals.

$6 + 2(15 - x) =$

$6 + (2 \times 15) + (2 \times -x) =$

$6 + 30 - 2x =$

$36 - 2x$

33) The correct answer is B. The total of the monthly payments is:

$700 per month × 360 months = $252,000

The total price of the house is $300,000 so deduct the total payments from this amount in order to calculate the immediate payment: $300,000 – $252,000 = $48,000

34) The correct answer is A. To divide, invert the second fraction and then multiply as shown.

$\frac{x}{5} \div \frac{9}{y} = \frac{x}{5} \times \frac{y}{9} = \frac{x \times y}{5 \times 9} = \frac{xy}{45}$

35) The correct answers are D and E. The formula for the area of a circle is: πR^2. The area of circle A is $\pi \times 5^2 = 25\pi$ and the area of circle B is $\pi \times 3^2 = 9\pi$. So, the difference between the areas is 16π.

The formula for circumference is: $\pi 2R$. The circumference of circle A is $\pi \times 2 \times 5 = 10\pi$ and the circumference for circle B is $\pi \times 2 \times 3 = 6\pi$. The difference in the circumferences is 4π. So, answer D is correct.

Diameter is equal to the radius times 2, so the diameter of circle A is 10 and the diameter of circle B is 6, and the difference in diameters is 4. So, answer E is also correct.

36) The correct answer is C. Calculate the area for each of the cupboards: 8 × 2 = 16 and 5 × 2 = 10. Therefore, the total area for both cupboards is 16 + 10 = 26. Then find the area for the entire kitchen: 8 × 12 = 96. Then deduct the cupboards from the total: 96 – 26 = 70

37) The correct answer is B. Circumference is $2\pi R$, so the circumference of the large wheel is 20π and the circumference of the smaller wheel is 12π. If the large wheel travels 360 revolutions, it travels a distance of: $20\pi \times 360 = 7200\pi$. To determine the number of revolutions the small wheel needs to make to go the same distance, we divide the distance by the circumference of the smaller wheel: $7200\pi \div 12\pi = 600$. Finally, calculate the difference in the number of revolutions: $600 - 360 = 240$

38) The correct answer is B. In June, Dawson County had 1.1 inches of rain and Emery County had 1.7 inches. Therefore, Emery County had 0.6 more inches of rainfall than Dawson County.

39) The correct answer is B. The two sides of the field form a right angle, so we can use the Pythagorean theorem to find the length of the hypotenuse to solve the problem: $\sqrt{3^2 + 4^2} = \sqrt{9 + 16} = \sqrt{25} = 5$

40) The correct answer is A. First, we need to find the circumference of the semicircle on the left side of the figure. The width of the rectangle of 10 inches forms the diameter of the semicircle, so the circumference of an entire circle with a diameter of 10 inches would be 10π inches. We need the circumference for a semicircle only, which is half of a circle, so we need to divide the circumference by 2: $10\pi \div 2 = 5\pi$. Since the right side of the figure is an equilateral triangle, the two sides of the triangle have the same length as the width of the rectangle, so they are 10 inches each. Finally, you need to add up the lengths of all of the sides to get the answer: 18 + 18 + 10 + 10 + 5π = 56 + 5π inches

41) The correct answer is B. We know that Paul will have run ten miles when he finishes since he runs 5 miles north, then returns and goes 5 miles south. The question is asking about his running pace or speed. In order to know speed, we need to know the distance traveled and the amount of time it takes to travel the distance. So, we know the distance, but not the time. Accordingly, we would need to know what time he gets back home in order to solve the problem.

42) The correct answer is A. You need to determine the death rate, so subtract the survival rate from 100% to get the death rate for each category. Then multiply for each category and compare:

Cardiopulmonary and vascular deaths: 602,000 × 0.18 = 108,360

HIV/AIDS deaths: 215,000 × 0.27 = 58,050

Diabetes deaths: 793,000 × 0.11 = 87,230

Cancer and leukemia deaths: 231,000 × 0.52 =120,120

Deaths from premature birth complications: 68,000 × 0.36 = 24,480

Then add the two smallest amounts together to solve: 24,480 + 58,050 = 82,530

43) The correct answer is a numerator value of 1 and a denominator value of 4. Calculate the area of the triangle: $\frac{1}{2} \times base \times height = \frac{1}{2} \times 5 \times 6 = \frac{1}{2} \times 30 = 15$. The height of the unshaded part is 9 inches since $15 - 6 = 9$, so next we can calculate the area of the unshaded rectangular part: $base \times height = 5 \times 9 = 45$. Add the area of the unshaded part of the figure to the area of the triangle in order to get the area for the entire figure: $45 + 15 = 60$. Finally, express the result as a simplified fraction with the area of the triangle in the numerator and the area of the entire figure in the denominator: $^{15}/_{60} = {}^{1}/_{4}$

44) The correct answer is A. We can see that December has the highest figure for all three of the lines. Accordingly, December will also have the greatest combined sales for all three companies.

45) The correct answers are B and C. When a transversal crosses two parallel lines, opposite angles will be equal in measure and corresponding angles will also be equal in measure. (Corresponding angles are angles in the matching same-shaped corners.) Angles ∠b and ∠c are opposite angles and angles ∠c and ∠f are corresponding angles, so answer B is correct. Angles ∠b and ∠e are corresponding and angles ∠e and ∠f are opposite, so answer C is also correct.

46) The correct answer is B. Our data set is: 2.5, 9.4, 3.1, 1.7, 3.2, 8.2, 4.5, 6.4, 7.8. First, put the numbers in ascending order: 1.7, 2.5, 3.1, 3.2, 4.5, 6.4, 7.8, 8.2, 9.4. The median is the number in the middle of the set: 1.7, 2.5, 3.1, 3.2, **4.5**, 6.4, 7.8, 8.2, 9.4

47) The correct answer is A. The outcome of an earlier roll does not affect the outcome of the next roll. When rolling a pair of dice, the possibility of an odd number is always $^{1}/_{2}$, just as the possibility of an even number is always $^{1}/_{2}$. We can prove this mathematically by looking at the possible outcomes:

1,1 1,2 1,3 1,4 1,5 1,6
2,1 2,2 2,3 2,4 2,5 2,6
3,1 3,2 3,3 3,4 3,5 3,6
4,1 4,2 4,3 4,4 4,5 4,6
5,1 5,2 5,3 5,4 5,5 5,6
6,1 6,2 6,3 6,4 6,5 6,6

The odd number combinations are highlighted:

1,1 **1,2** 1,3 **1,4** 1,5 **1,6**
2,1 2,2 **2,3** 2,4 **2,5** 2,6
3,1 **3,2** 3,3 **3,4** 3,5 **3,6**
4,1 4,2 **4,3** 4,4 **4,5** 4,6
5,1 **5,2** 5,3 **5,4** 5,5 **5,6**
6,1 6,2 **6,3** 6,4 **6,5** 6,6

So, we can see that an odd number will be rolled half of the time.

48) The correct answer is E. The proposed course of action affects the residents of Buford, so the statistical sample must represent these residents as much as possible. The most representative sample would therefore be achieved by selecting participants for a survey from a list of all of the citizens living in Buford. Answers A and B may represent people living in other areas who are merely driving near Buford. Answers C and D would fail to represent residents of Buford who do not drive or who have not registered to vote.

49) The correct answer is D. To find the mean, add up all of the items in the set and then divide by the number of items in the set. Here we have 7 numbers in the set, so we get our answer as follows: (89 + 65 + 75 + 68 + 82 + 74 + 86) ÷ 7 = 539 ÷ 7 = 77

50) The correct answer is B. At the start of the party, there are 10 red balloons, 7 green balloons, 6 purple balloons, 5 orange balloons, and 11 blue balloons in the bag, so we add all of these up to get our data set at the beginning: 10 + 7 + 6 + 5 + 11 = 39 items in the data set at the beginning. Then a blue balloon and an orange balloon are removed, so we need to reduce the data set for these two items: 39 – 2 = 37. The problem is asking about the probability of a blue balloon. There are 11 blue balloons at the start and one has been removed, so there are 10 blue balloons left. Remember to express the probability as a fraction with the possibility of the outcome in the numerator and the remaining data set in the denominator: $^{10}/_{37}$

51) The correct answer is E. We have the data set: 1.6, 2.9, 4.5, 2.5, 2.6, 5.1, 5.4 The mode is the number that occurs most frequently. All of the numbers only occur once. So, there is no mode.

52) The correct answer is C. Swap the digits on each of the numbers and subtract to find the one that is off by 36. In doing so, we discover that Sam entered 95 when he should have entered 59, because the difference is 36: 95 – 59 = 36.

53) The correct answer is B. We don't know the age of the 10th car, so put this in as x to solve: (2 + 3 + 4 + 5 + 6 + 7 + 9 + 10 + 12 + x) ÷ 10 = 6

[(2 + 3 + 4 + 5 + 6 + 7 + 9 + 10 + 12 + x) ÷ 10] × 10 = 6 × 10

2 + 3 + 4 + 5 + 6 + 7 + 9 + 10 + 12 + x = 60

58 + x = 60

x = 2

54) The correct answers are (2) and (5). Statistical questions ask about behaviors or opinions. So, the following are statistical questions: "How many residents of the town oppose the tax increase?" and "Do most college graduates in our state find jobs within one year of leaving college?"

55) The correct answers are A and B. The range is the difference between the highest number and the lowest number in a data set. If the range increases by 2, then the highest number could go up by 2 or the highest and lowest numbers could increase and decrease by 1 each respectively.

56) The correct answer is E. We have 54 cards in the deck (13 × 4 = 52). We have taken out two spades, one heart, and a club, thereby removing 4 cards. So, the available data set is 48 (52 – 4 = 48). The desired outcome is drawing a heart. We have 13 hearts to begin with and one has been removed, so there are 12 hearts left. So, the probability of drawing a heart is $^{12}/_{48} = {^1}/_4$

Math Formula Sheet

Distance:
1 foot = 12 inches
1 yard = 3 feet
1 meter = 1,000 millimeters
1 meter = 100 centimeters

Rectangle and box perimeter:
perimeter = 2(*length* + *width*)

Area of squares and rectangles:
area = *length* × *width*

Volume of rectangular or square solid:
Length × Width × Height

Circles:
number of degrees in circle = 360°
circumference = π × *diameter*
area = π × (*radius*)^2
$\pi \approx 3.14$

Triangles:
sum of angles = 180°
area = ½ (*base* × *height*)
hypotenuse length: $\sqrt{A^2 + B^2} = C$
The hypotenuse is the diagonal side of a triangle, opposite the right angle.

Straight Lines:
sum of two supplementary angles = 180°
Note that supplementary angles are side-by-side on a straight line.

Praxis Core Reading Test Format

The Praxis Core Reading Test (5713) contains the following types of questions:

- Identifying details and key ideas – 17 to 22 questions
- Understanding the writer's craft and language – 14 to 19 questions
- Evaluating relationships, data, and evidence – 17 to 22 questions

There are 56 questions in total on the Praxis Core Reading Exam.

You will take the exam on a computer, unless you have applied for an exemption.

You will have 85 minutes to take the Praxis Core Reading Test.

You will see a combination of the following types of reading texts on the Praxis Core Reading Test (5713):

- Paired passages consisting of approximately 200 words each
- Long passages of 200 words
- Shorter passages consisting of approximately 100 words
- Brief statements

Identifying details and key ideas questions require you to understand:

- the main idea
- specific details
- inferences
- the basis of the author's argument

Understanding the writer's craft and language questions will ask you to evaluate and interpret:

- the author's tone and purpose
- the author's viewpoint on the subject
- the structure of the text
- facts and opinions
- the use of language, including figurative language and word meanings

Evaluating relationships, data, and evidence questions cover the following skills:

- analyzing data in charts and graphs
- understanding the relationships between claims in texts
- determining whether evidence strengthens or weakens an argument
- identifying the author's assumptions
- drawing conclusions based on the information provided

For the computer-delivered version of the test, questions 1 to 19 will require you to click on the part of the sentence that contains an error.

For the remaining questions on the reading test, you may need to respond by:

- clicking on an oval
- typing words into an on-screen box
- checking a box on the screen
- clicking on part of a graphic
- selecting options from a menu

With every question that the computer delivers, you will see instructions on the screen.

Praxis Core Reading Practice Test 1

Instructions: After reading each passage, choose the best answers to the questions that follow.

Question 1 refers to the following passage.

"All knowledge that is about human society, and not about the natural world, is historical knowledge, and therefore rests upon judgment and interpretation. This is not to say that facts or data are non-existent, but that facts get their importance from what is made of them in interpretation, for interpretations depend very much on who the interpreter is, who he or she is addressing, what his or her purpose is, and at what historical moment the interpretation takes place" (Excerpt from *Culture and Imperialism,* Edward Said).

1. The primary purpose of the passage is to:
   - A. assert that historical knowledge diverges from knowledge about nature.
   - B. emphasize the historical significance of facts and data.
   - C. argue that historical knowledge hinges on analyses and opinions.
   - D. point out that historical knowledge is dubious from an academic perspective.
   - E. explain the way in which historical interpretations are made.

Question 2 refers to the following passage.

"Celebrity" is the term used to describe someone who is famous and attracts attention from the general public and the world's media. Traditionally, a celebrity would gain the title by his or her work or achievements in a particular field of expertise. Actors, musicians, politicians, and inventors have all become celebrities in the past. However, as the twenty-first century progresses, a new celebrity has arrived – the nobody. As one peruses glossy TV magazines, it is easy to notice the amount of reality shows that now dominate our screens – Wife Swap, American Idol, America's Got Talent, and the reality pioneer Big Brother. The concept itself of Big Brother is everything that George Orwell warned us about: "normal" people are thrust into the limelight to be mocked, glorified, vilified, and humiliated in equal measures. And we lap it up.

2. The writer would most strongly agree with which one of the following statements?
   - A. Reality TV participants are so-called celebrities who have no real achievements or expertise.
   - B. Reality TV participants are foolish for wanting to appear on television.
   - C. The general public needs to stop watching reality TV shows in order to prevent the phenomenon of spurious celebrity.
   - D. Glossy TV magazines should stop promoting reality TV shows.
   - E. Big Brother is worse than the other reality shows mentioned in the passage.

Question 3 refers to the following passage.

For any state to make sex a qualification that must ever result in the disfranchisement of one entire half of the people is a violation of the supreme law of the land. By it, the blessings of liberty are forever withheld from women and their female posterity. To them, this government has no just powers derived from the consent of the governed. To them, this government is not a democracy. It is not a republic. It is an *odious* aristocracy; a hateful *oligarchy* of sex; the oligarchs over the mother and sisters, the wife and daughters, of every household – which ordains all men *sovereigns*, all women subjects, carries dissension, discord, and rebellion into every home of the nation (Excerpt from "On Women's Right to Vote" by Susan B. Anthony).

3. The use of the words "odious," "oligarchy," and "sovereigns" serves to support the argument that:
   - A. the entire populace is disenfranchised.
   - B. the government is too powerful.
   - C. the situation is unfair and change is necessary.
   - D. dissension and rebellion are bound to increase.
   - E. proposed changes are bound to be ineffectual.

Questions 4 to 8 refer to the following passage.

Working in a run-down laboratory near Paris, Marie Curie worked around the clock to discover a radioactive element. When she finally captured her *quarry*, she named it “radium” after the Latin word meaning ray. She had spent the day blending chemical compounds which could be used to destroy unhealthy cells in the body. As she was about to retire to bed that evening, she decided to return to her lab. There she found that the chemical compound had become crystalized in the bowls and was emitting the elusive light that she sought.

Inspired by the French scientist Henri Becquerel, Curie won the Nobel Prize for Chemistry for her discovery. Upon winning the prize, she declared that the radioactive element would be used only to treat disease and would not be used for commercial profit. Today radium provides an effective remedy for certain types of cancer. Radium, now used for a treatment called radiotherapy, works by inundating diseased cells with radioactive particles. Its success lies in the fact that it eradicates malignant cells without any lasting ill effects on the body.

4. Which of the following best states the main idea of the passage?
   A. Marie Curie was an inventive and enterprising chemist.
   B. Marie Curie made many important discoveries during her illustrious career.
   C. Radium is effective in treating cancer because it inundates diseased cells with radioactive particles.
   D. Henri Becquerel was an inspiration to Marie Curie, and his influence is notable in her work.
   E. Marie Curie worked assiduously to discover radium and insisted that the element should only be used for medical purposes.

5. As used in sentence 2, “quarry” most nearly means:
   A. a precious commodity
   B. an unknown catalyst
   C. a chemical compound
   D. an object that is sought
   E. a source that emits light

6. According to the passage, which of the following is the reason why radium is used as a cancer treatment?
   A. because it is cost effective
   B. because it destroys cancerous cells
   C. because it has no long-term effects
   D. because it emits a glowing light
   E. because it derives from a radioactive element

7. In paragraph 2, the author notes that “Curie won the Nobel Prize for Chemistry for her discovery” in order to emphasize:
   A. the importance of her achievement.
   B. that the Nobel Prize is not usually awarded for chemistry.
   C. Curie’s background as a chemist.
   D. that it may have been controversial to award the prize to a woman.
   E. the amount of commercial profit that was lost in making the discovery.

8. Which of the following best describes the organization of the passage?
   A. A lingering doubt is clarified, and then a further question is posed.
   B. The background to a discovery is discussed, and then the reasons for its current use are explained.
   C. A problem is presented, and then a solution is proposed.
   D. Historical information is provided, and then a unique scenario is presented to illustrate the information.
   E. A scientific phenomenon is summarized, and then a possible controversy is mentioned.

Questions 9 to 12 refer to the following passage.

Michelangelo began work on the massive project of painting of the ceiling of the Sistine Chapel in Italy in the summer of 1508, assisted by six others who helped to mix his paint and plaster. However, as work proceeded, the artist dismissed each of his assistants one by one, claiming that they lacked the competence necessary to do the task at hand. Described as the lonely genius, the painter himself often felt incompetent to complete the project entrusted to him by Pope Julius II. Having trained as a sculptor, Michelangelo had an extremely low opinion of his own painting skills. Yet, he went on to paint one of the most beautiful works in art history. In spite of his frequent personal misgivings, he persevered to paint the ceiling with his vision of the creation of the universe. The nine series of scenes that he painted include the Separation of Light from Darkness, the Drunkenness of Noah, the Ancestors of Christ, and the Salvation of Mankind.

9. The author suggests that which of the following probably occurred after Michelangelo dismissed his assistants?
   A. It took the artist much longer to finish the project than expected.
   B. The public's opinion of the artist and his work began to deteriorate.
   C. The artist wished that he could go back to his career as a sculptor.
   D. The artist sometimes felt isolated and overwhelmed by the project.
   E. The Pope had misgivings about having entrusted the project to the artist.

10. The author mentions Pope Julius II most likely in order to:
   A. provide an exception to a general rule.
   B. explain the customs and manners of the past.
   C. support his assertion about the historical importance of the work of art.
   D. imply that the artist did not have freedom of expression.
   E. contrast artistic works of the past with those of modern times.

11. It can be inferred that the author views Michelangelo's painting of the Sistine Chapel as:
   A. the most esthetically-pleasing painting in all of Italy.
   B. the first example of art portraying a series of different scenes.
   C. a representative example of the art of that era.
   D. a typical example of Italian art.
   E. an artistic treasure that has endured for centuries.

12. In the last sentence of the passage, the author mentions that nine scenes were painted on the ceiling of the Sistine Chapel to exemplify:
   A. the artist's personal misgivings.
   B. the size of the chapel.
   C. the enormity of the project.
   D. the incompetence of the artist's assistants.
   E. how paint and plaster can be mixed to paint art on a ceiling.

Questions 13 to 15 refer to the following passage.

Highly concentrated radioactive waste is lethal and can remain so for thousands of years. Accordingly, the disposal of this material remains an issue in most energy-producing countries around the world. In the United States, for example, liquid forms of radioactive waste are usually stored in stainless steel tanks. For extra protection, the tanks are double-walled and surrounded by a concrete covering that is one meter thick. This storage solution is also utilized the United Kingdom, in most cases.

The potential future problem lies in the fact that nuclear waste generates heat as radioactive atoms decay, thereby creating a high risk of a radioactive leak. Therefore, the liquid needs to be cooled by pumping cold water into coils inside the tanks. However, *this strategy* is only a temporary storage solution. The answer to the long-term storage of nuclear waste may be fusing the waste into glass cylinders that are stored deep underground.

13. The passage indicates that which of the following helps to prevent leaks from the tanks used for storing radioactive waste?
   A. The tanks are encased in concrete.
   B. The tanks only contain waste in liquid form.
   C. The tanks provide a place where radioactive atoms can decay.
   D. The tanks are combined with cold water.
   E. The tanks are fused into glass cylinders.

14. In the second paragraph of the passage, "this strategy" refers to which of the following?
   A. the danger of the consequences of the escape of radioactive substances from the storage tanks
   B. the generation of nuclear waste as radioactive atoms decay
   C. the storage of the waste in double-walled tanks
   D. the fact that the solution is only viable on a short-term basis
   E. the process of cooling the liquid by means of cold water

15. The author's attitude can best be described as one of:
   A. apprehension
   B. disillusionment
   C. shock
   D. contempt
   E. appreciation

Questions 16 to 18 refer to the following passage.

Equating the whole history of the struggle of humankind to that of the class struggle, the social and political writings of Karl Marx have been the impetus of a great deal of change within society. According to Marxism, the political school of thought based on Marx's doctrines, the working class should strive to defeat capitalism, since capitalistic societies inherently have within them a dynamic that results in the wealthy ruling classes oppressing the masses. The nation state is seen as an instrument of class rule because it supports private capital and suppresses the common person through economic mechanisms, such as the taxation of wages. Because growth of private capital is stimulated by earning profits and extracting surplus value in the production process, wages have to be kept low.

Since capitalism reduces the purchasing power of workers to consume the goods that they produce, Marx emphasized that capitalism inheres in a central contradiction. Under the tenets of Marxism, capitalism is therefore inherently unstable. Marx asserted that productive power ideally should be in the hands of the general public, which would cause class differences to vanish. These idealistic writings have had a huge impact on culture and politics; yet, many believe that Marx's work lacked the practical details needed to bring about the changes to the class structure that he envisaged.

16. The primary purpose of the first paragraph is to:
   A. discuss why the writings of Karl Marx have had such enduring social and political importance.
   B. explain the basic tenets of Marxism, before going on to discuss Marxist views on capitalism and the consequences of private capital.
   C. critique the existing class structure and oppression of the masses.
   D. decry the manner in which taxation and earning profits cause wages to be lowered.
   E. recount the cultural and historical significance of the work of Karl Marx.

17. Which of the following best describes the relationship between the two paragraphs in the passage?
   A. The first paragraph states an assertion, while the second paragraph refutes that assertion with statistical evidence.
   B. The first paragraph explains a long-standing problem, and the second paragraph provides the potential solution.

C. The first paragraph introduces and expounds upon a theory, while the second paragraph points out criticisms of the theory.
D. The first paragraph gives the background to the topic in a general way, and the second paragraph provides specific details about the topic.
E. The second paragraph provides further examples of the phenomenon mentioned in the first paragraph.

18. The writer mentions the "huge impact" that these writings have had on culture and politics in the last sentence in order to:
A. underscore the fact that class differences have not yet vanished.
B. highlight the way in which capitalism is often unstable.
C. reiterate the importance of giving power back to the general populace.
D. lament the social change that Marx himself predicted.
E. juxtapose this impact to Marx's failure to include pragmatic instructions in his work.

Question 19 refers to the following passage.

The notion of liberal arts education is believed to have been established in ancient Greece. Including the disciplines of logic, rhetoric, and grammar, a liberal arts education in those days was designed to train members of society to undertake important civic duties, such as jury service and public debate. In modern parlance, the term "liberal arts education" can be interpreted in a variety of ways, although it is generally taken to mean that the studies will include courses in one or more of the subject areas of the humanities, such as languages, literature, or philosophy.

19. The passage provides information for most fully answering which of the following questions?
A. Why was liberal arts education originally established in Greece?
B. How did students study logic, rhetoric, and grammar in ancient times?
C. Why were jury service and public debate considered to be important civic duties?
D. What is the background to and definition of the term "liberal arts education"?
E. Why do the humanities include languages, literature, and philosophy?

Question 20 refers to the following passage.

We stand today at the threshold of a great event both in the life of the United Nations and in the life of mankind, that is the approval by the General Assembly of the Declaration of Human Rights. This declaration may well become the international Magna Carta of all people everywhere. We hope its proclamation by the General Assembly will be an event comparable to the proclamation of the Declaration of the Rights of Man by the French people in 1789, the adoption of the Bill of Rights by the people of the United States, and the adoption of comparable declarations at different times in other countries (Excerpt from "Adoption of the Declaration of Human Rights" by Eleanor Roosevelt).

20. The speaker most likely mentions the Magna Carta, the Declaration of the Rights of Man, and the Bill of Rights in order to:
A. incite dissent among the audience members.
B. emphasize the historical importance of this event.
C. persuade her opponents to support this declaration.
D. predict future proclamations by the General Assembly.
E. perpetuate the status quo.

Questions 21 to 23 refer to the following passage.

The world's first public railway carried passengers, even though it was primarily designed to transport coal from inland mines to ports on the North Sea. Unveiled on September 27, 1825, the train had 32 open wagons and carried over 300 people. The locomotive steam engine was powered by what was termed the steam-blast technique. The chimney of the locomotive redirected exhaust steam into the engine via a narrow pipe. In this way, the steam created a draft of air which followed after it, creating more power and speed for the engine.

The train had rimmed wheels which ran atop rails that were specially designed to give the carriages a faster and smoother ride. While the small carriages could hardly be termed *commodious*, the locomotive could accelerate to 15 miles per hour, a record-breaking speed at that time. Subsequently, the inventor of the locomotive, George Stephenson, revolutionized his steam engine by adding 24 further pipes. Now containing 25 tubes instead of one, Stephenson's second "iron horse" was even faster and more powerful than his first creation.

21. The author most probably uses the word "commodious" in order to mean:
    A. small
    B. uncomfortable
    C. spacious
    D. speedy
    E. smooth

22. Why was the second locomotive that Stephenson invented an improvement on his first?
    A. because it ran with greater force and speed
    B. because it was more comfortable
    C. because it could carry more passengers
    D. because it contained more pipes and tubes
    E. because it ran more smoothly

23. From the information contained in the passage, it seems reasonable to infer which of the following?
    A. Many passengers were frightened about traveling on Stephenson's new locomotive.
    B. George Stephenson's inventions laid the basic foundations for modern day public trains and railways.
    C. Profits in the coal industry increased after the invention of the locomotive.
    D. Stephenson should have been able to invent a locomotive that could run faster.
    E. Stephenson's second locomotive carried more passengers than his first one.

Question 24 refers to the following passage.

Although there are many different types and sizes of coins in various countries, vending machines around the world operate on the same basic principles. The first check is the slot: coins that are bent or too large will not go in. Once inside the machine, coins fall into a cradle which weighs them. If a coin is too light, it is rejected and returned to the customer. Coins that pass the weight test are then passed along a runway beside a magnet. Electricity passes through the magnet, causing the coin to slow down in some cases. If the coin begins to slow down, its metallurgic composition has been deemed to be correct. The coin's slow speed causes it to miss the next obstacle, the deflector. Instead, the coin falls into the "accept" channel, and the customer receives the product.

24. According to the passage, how is the metallurgical composition of a coin determined to be correct?
    A. By its weight
    B. By its increased velocity in the runway
    C. By whether it runs alongside the magnet
    D. By the electricity that has passed through the magnet
    E. By missing the deflector

Questions 25 to 28 refer to the following passage.

In December of 1880, a friend who was a veterinary surgeon gave Louis Pasteur two rabid dogs for research purposes. Victims of bites from rabid dogs normally showed no symptoms for three to twelve weeks. By then, however, the patient would be suffering from convulsions and delirium, and it would be too late to administer any remedy.

So-called treatments at that time consisted of burning the bitten area of skin with red-hot pokers or with carbolic acid. Pasteur devoted himself to discovering a more humane and effective method of treatment for the disease. His tests on the rabid dogs confirmed that rabies germs were isolated in the saliva and nervous systems of the animals. After many weeks of tests and experiments, Pasteur cultivated a vaccine. Derived from a weakened form of the rabies virus itself, the vaccine is administered before the microorganism is encountered and stimulates the immune system to recognize and fight off any future exposure to the organism.

25. In can be inferred from the passage that patients today would most likely respond to the prospect of treatments that were used in the past with:
    A. fear
    B. bewilderment
    C. scorn
    D. apathy
    E. resignation

26. The primary purpose of the passage is to discuss:
    A. pasteurization and the rabies vaccine.
    B. the life and work of Louis Pasteur.
    C. Pasteur's discovery of the rabies vaccine.
    D. experimental research on rabid dogs.
    E. uses of the modern rabies vaccine.

27. The passage suggests that the discovery of the rabies vaccine was significant for which of the following reasons?
    A. It prevented animals from suffering during scientific experiments.
    B. It led to advancements in other areas of medical science.
    C. It caused the treatments for other diseases to become more humane and effective.
    D. It contributed to the prevention of the contagions in germs, in general.
    E. It helped many people avoid physical suffering and death.

28. Which of the following situations provides the closest example to the use of a vaccine, as it is described in the passage?
    A. a patient who is given an antibiotic to recover from an infection
    B. a person who gargles every day in the belief that it will help prevent catching a cold
    C. a person who takes medicine after having been told she has high cholesterol
    D. children who get injections to prevent catching mumps and measles
    E. the use of anesthetic to put patients asleep during operations

Questions 29 to 33 refer to the following pair of passages.

Passage 1
Abraham Lincoln observed that happiness is a choice for most people. This echoes the claims of the Dalai Lama, who stated that people can decide whether they will be happy or not through self-discipline. So, isn't the choice simple really? Shouldn't we choose to be happy? Being happy occurs when we choose not to worry. This choice is based on a thankful attitude. We have so much to be thankful for. Thank the taxi driver for bringing you home safely, thank the cook for a wonderful dinner, and thank the person who cleans your windows. When we give thanks to others whenever possible, we choose the path of gratitude that leads to the road to happiness.

Passage 2
Almost everyone has heard the hit single "Don't Worry, Be Happy" by Bobby McFerrin. But isn't McFerrin's refrain that everyone can choose to be happy by simply deciding not to worry overly simplistic in reality? Living a happy and worry-free life is a wonderful ideal, but it must be said that life is full of stresses and strains that are often not of our own choosing. One of the truest things ever said is that the only thing in life that will always remain the same is change. In addition to causing us to worry, stress is linked to the top causes of death, such as heart disease, cancer, and stroke. So, achieving happiness in today's society is often a complex, multi-dimensional process.

29. Which of the following statements best describes the relationship between Passage 1 and Passage 2?
    A. Passage 1 introduces a theory that is supported by the examples in Passage 2.
    B. Passage 1 explains a philosophy that conflicts with the personal anecdote recounted in Passage 2.
    C. Passage 1 describes a problematic situation that is ameliorated by the remedy described in Passage 2.
    D. Passage 2 provides an argument that undermines the viewpoint expressed in Passage 1.
    E. Passage 2 challenges the evidence that is introduced in Passage 1.

30. With which of the following ideas do both authors agree?
    A. Happiness is the result of decisions we make.
    B. Most people are not thankful enough.
    C. Many people lack of self-discipline.
    D. Change is the one constant thing in life.
    E. Worry has a direct impact on happiness.

31. The use of the phrase "choose to be happy" in both passages serves to:
    A. convey two differing opinions on whether a happy, worry-free life can really be achieved.
    B. emphasize the view that the decisions that people make have the greatest influence on their lives.
    C. demonstrate both authors' support for the idea that happiness is a choice.
    D. illustrate the disagreement over how people should express their gratitude.
    E. reveal that achieving happiness can be a complex process.

32. How would the writer of Passage 1 most likely respond to the following statement from Passage 2?: "It must be said that life is full of stresses and strains that are often not of our own choosing."
    A. People who experience stress are ungrateful for the positive aspects of their lives.
    B. Abraham Lincoln and the Dalai Lama also experienced stress.
    C. It may be true that we cannot choose certain events in life, but we can still choose to be happy.
    D. Stressed-out people lack self-discipline generally.
    E. Stress occurs only because people worry too much.

33. Both authors do which of the following?
    A. Acknowledge recent trends that affect the subject
    B. Cite the names of authorities on the subject
    C. Pose rhetorical questions to the reader
    D. Provide explanations of technical language
    E. Recount the minutiae of a personal experience

Question 34 refers to the following passage.

There has been a fundamental change in the relationship between the actor and the audience in recent years. According to Aristotelian principles, actors should provoke an emotional catharsis in the members of the audience. Traditionally, actors have provided this emotional release, but this is far from the case in the performances in many modern motion pictures and television programs. Even though many productions are increasingly based on contrived stories or weak plots, modern actors could do more with these roles than merely create mindless diversions. Sadly, many performers nowadays lack gravitas. When actors engage in such vacuous performances, they do not even begin to serve the higher purpose of their profession.

34. According to the passage, one way to distinguish a traditional actor from a modern one is by considering:
   A. the seriousness of the subject matter of the production.
   B. the emotional response of the viewers.
   C. the strength of the plot.
   D. the originality of the storyline.
   E. whether the performance grabs the attention of the audience.

Question 35 refers to the following passage.

This line of inquiry is wholly improper and argumentative. It is not a statement as to what the issues are. Your Honor has already held that this act is constitutional, it being the law of the land. There is but one issue before this court and jury, and that is, whether the defendant violated the statute. If Your Honor would please consider that some of the witnesses in this case are not very well and others are awfully ignorant. Furthermore, we have just agreed among ourselves to disregard the so-called evidence and argue the case (Excerpt from the *State v. Scopes Trial*, Delivered by Ben G. McKenzie).

35. The speaker's attitude toward the witnesses is best described as one of:
   A. disrespect
   B. equanimity
   C. impartiality
   D. understanding
   E. despair

Question 36 refers to the following passage.

According to Stephen Krashen's input hypothesis, a language learner improves his or her language skills when he or she is exposed to language input such as lectures or reading materials that are one level above the learner's current level of language ability. Language output such as verbal or written expressions are not seen to have any direct effect of the learner's ability.

36. According to the passage, which of the following is the best definition of the input hypothesis?
   A. It is an assumption that all language learners begin at the same level of ability.
   B. It is a theory which asserts that learners can best improve their language skills when their learning is appropriately challenging.
   C. It is a school of thought that discounts the importance of traditional grammatical skills.
   D. It is a system of language rules established by Stephen Krashen that learners of new languages try to follow.
   E. It is an educational methodology devised by language teachers.

Questions 37 to 39 refer to the following passage.

The pyramids at Giza in Egypt are still among the world's largest structures, even today. The monuments were constructed well before the wheel was invented, and it is notable that the Egyptians had only the most primitive, handmade tools to complete the massive project. Copper saws were used to cut softer stones, as well as the large wooden posts that levered the stone blocks into their final places. Wooden mallets were used to drive flint wedges into rocks in order to split them. The Egyptians also utilized drills that were fashioned from wood and twine. In order to ensure that the stones were level, wooden rods were joined by strips of twine to check that the surfaces of the stone blocks were flat. Finally, the stone blocks were put onto wooden rockers so that they could more easily be placed into their correct positions on the pyramid.

37. According to the passage, the two tools which were used to place the stones into their final positions on the pyramid were made from which substance?
A. flint
B. copper
C. twine
D. stone
E. wood

38. What is the writer's main purpose?
A. to give a step-by-step explanation of the construction of the Giza pyramids
B. to compare the construction of the Giza pyramids to that of modern-day structures
C. to give an overview of some of the main implements that were used to construct the Giza pyramids
D. to highlight the importance of the achievement of the construction of the Giza pyramids
E. to bring to light a misconception in previous accounts of the construction of the Giza pyramids

39. The passage suggests that the Giza pyramids are significant for which of the following reasons?
A. It is incredible that the Egyptians were able to construct the pyramids using only hand-made tools.
B. It is a pity that the wheel was not available to the Egyptians during the construction of the pyramids at Giza.
C. Modern construction projects could learn from the example of the Giza pyramids.
D. The most difficult aspect of the project was placing the stones in the correct position on the pyramid.
E. The pyramids could have been larger if modern tools had been available.

Questions 40 to 42 refer to the following passage.

Recent research shows that social media platforms may actually be making us antisocial. Survey results indicate that many people would prefer to interact on Facebook or Twitter, rather than see friends and family in person. The primary reason cited for this phenomenon was that one does not need to go to the effort to dress up and travel in order to use these social media platforms.

Another independent survey revealed that people often remain glued to their hand-held devices to check their social media when they do go out with friends. It seems that social media platforms may be having a detrimental impact upon our social skills and interpersonal relationships.

40. In the second paragraph, the author suggests that the effect of social media on relationships has largely been:
A. deleterious
B. ambivalent
C. ambiguous
D. disinterested
E. objective

41. The author would most likely recommend that users of social media do which of the following?
A. go out more often
B. obtain a wider circle of friends
C. see family members only in person
D. turn off their hand-held devices when they are out with friends
E. limit their use of Facebook to evening hours

42. Which of the following, if true, would best support the argument presented in the passage?
A. Recent research indicates that there has been a rise in antisocial behavior in public places.
B. Statistics reveal that most children would be reluctant to visit relatives who do not allow the use of smart phones while at the table.
C. People who feel addicted to their electronic devices may require psychological counseling.

D. Most people state that they are in control of when and how they use their electronic devices.
E. Most children would rather be outdoors than watch television.

Questions 43 to 45 refer to the following information.

Nassir has a busy work schedule and an active social life. His normal work schedule every week is from Monday to Friday, 8 am to 5 pm. Although it is sometimes necessary, he prefers not to work after 5:00 pm as he likes to visit with friends or take part in his hobbies during the evening. A chart of Nassir's commitments for the month of February is shown below.

Sunday	Monday	Tuesday	Wednesday	Thursday	Friday	Saturday
1	2	3 6:00 pm – personal training at gym	4 7:00 pm – meet Ali in town	5 Staff meeting at work – 1:30 pm	6 8:00 pm – Martial arts practice	7
8	9 Work from 1:00 pm to 9:00 pm this week	10 8:30 am – personal training at gym	11 Staff meeting at work – 3:30 pm	12 9:30 am – personal training at gym	13 10:00 am – meet Marta for brunch	14 8:00 am – Go on hike with Martin and Suki
15 5:00 pm – meet Ali in town	16 8:00 pm – meet Terri for a late supper	17 7:00 pm – Martial arts practice	18 8:30 pm – personal training at gym	19 7:30 pm – event at public library	20 6:00 pm – meet Jacinta for coffee	21 Go to Mom and Dad's this weekend
22 2:00 pm – meet Ali in town	23 Work from 1:00 pm to 5:00 pm this week	24 6:30 pm – meet Tom for supper	25 6:00 pm – meet Jacinta for coffee	26 Staff meeting at work – 2:00 pm	27 8:00 am – Martial arts practice	28 10:00 am – Meet Jim for bike ride

43. Which of the following statements is best supported by the information provided above?
A. Nassir has staff meetings once a week.
B. Nassir normally sees his parents once a month.
C. Nassir finds it difficult to work late in the evening.
D. Nassir never has to work on Saturday or Sunday.
E. Nassir writes his work hours on his calendar only when they vary from 8 am to 5 pm.

44. Based on the information provided above, which of the following statements best explains how Nassir plans the time that he spends exercising?
A. Nassir's gym is open 24 hours a day, so he can go there any time.
B. Nassir can go to martial arts practice during the evening only.
C. Nassir does not like to exercise early in the morning.
D. Nassir likes to exercise with friends when possible.
E. Nassir never exercises with Ali in town.

45. During the month of February, which of the following activities did Nassir take part in at least four times? You may choose more than one answer.
A. Meeting friends for supper.
B. Meeting friends for coffee or brunch.
C. Meeting friends in town.
D. Personal training at the gym.
E. Martial arts training.

Questions 46 to 50 refer to the following information.

How is civil order maintained within any given population? The civil order control function suggests that public order is best maintained through agencies other than the police force or militia. Accordingly, martial law, the establishment of military rule over a civilian population, is only imposed when other methods of civil control have proven ineffective. Either the leader of a country's military system or of the country's own government may lay down the edict for the rule of martial law. In the past, this state of affairs most commonly occurred to quell uprisings during periods of colonial occupation or to thwart a coup, defined as an illegal and usually violent seizure of a government by a select group of individuals.

So, how is the declaration of martial law currently regulated? The constitutions of many countries now make provisions for the introduction of martial law, allowing it only in cases of national emergency or in the case of threats to national security from foreign countries. In democratic nations, severe restrictions are imposed on the implementation of martial law, meaning that a formal declaration of military rule over a nation should be rendered virtually impractical. In spite of these democratic systems being in place, forms of military control are still instituted during times of crisis, with a country's military system being mobilized to support civil authorities, such as municipalities and local police forces. The United States Secretary of State recently commented: "The use of military force to control the population is still a necessary albeit inimical outcome for the governments of certain countries around the globe today."

46. Which of the following statements best explains the differences between how martial law was instituted in the past and how it is instituted at present?
A. In the past, the militia was not used to support civil authorities, although it is used this way at present.
B. In the past, countries did not have constitutions or other established means to regulate the declaration of martial law.
C. There are more threats to national security nowadays than there were in the past.
D. Civil order was more difficult to maintain in the past than it is during the present time.
E. The Secretary of State did not have to make declarations of martial law in the past.

47. It can be inferred from the passage that the United States Secretary of State would agree with which of the following statements?
A. The declaration of martial law is sometimes needed, although it is usually undesirable.
B. The declaration of martial law is a pragmatic remedial solution when a population is out of control.
C. The country's military system should provide more support for civil authorities.
D. The police forces of most municipalities are already over-burdened with other tasks.
E. Martial law should automatically be established during times of national crisis.

48. The last sentence of the first paragraph suggests that which of the following is true of coups?
A. They usually represent a large proportion of the population.
B. They no longer occur as countries now have controls in place to prevent them.
C. They may involve assassinating or harming government leaders, officials, or citizens.
D. They have taken place most frequently during periods of colonial occupation.
E. Controlling coups is virtually impractical nowadays.

49. The author finds fault with the civil order control function for its failure to answer which of the following questions?
   A. How is civil order maintained in democratic nations?
   B. How is the declaration of martial law regulated at present?
   C. Why did uprisings take place in colonial times?
   D. What constitutional measures exist to regulate the declaration of martial law?
   E. Why do democratic nations sometimes deploy the military to impose order on their populations?

50. It can be inferred that the author of the passage would agree with the basic tenet of the civil order control function for which of the following reasons?
   A. Police forces are well trained and can readily respond during times of crisis.
   B. Public order and civil control are not as important as other social issues.
   C. A country's military can best control the civilian population.
   D. There are occasions when public order can only be reinstated through the establishment of military rule, in spite of the disadvantages in doing so.
   E. The leader of a country's military system usually works closely with the police force.

Questions 51 to 53 refer to the following information.

An efficient electron microscope can magnify an object by more than one million times its original size. This innovation has thereby allowed scientists to study the precise molecules that constitute human life. The electron microscope functions by emitting a stream of electrons from a gun-type instrument, which is similar to the *apparatus* used in an old-fashioned television tube. The electrons pass through an advanced electronic field that is accelerated to millions of volts in certain cases. Before traveling through a vacuum in order to remove oxygen molecules, the electrons are focused into a beam by way of magnetic coils. Invisible to the naked eye, electron beams can nevertheless be projected onto a florescent screen. When striking the screen, the electrons glow and can even be recorded on film.

In the transmission electron microscope, which is used to study cells or tissues, the beam passes through a thin slice of the specimen that is being studied. On the other hand, in the scanning electron microscope, which is used for tasks such as examining bullets and fibers, the beam is reflected. This reflection creates a picture of the specimen line by line.

51. What is the last step in the process by which the beam emanating from the electron microscope is formed?
   A. The electrons pass through an electronic field.
   B. The electrons are accelerated to millions of volts.
   C. The electrons travel through a vacuum.
   D. Oxygen is removed from the molecules.
   E. The electrons pass through magnetic coils.

52. Which of the following words, if substituted for the word "apparatus" in paragraph 1, would introduce the least change to the meaning of the sentence?
   A. machine
   B. electricity
   C. device
   D. tube
   E. bulb

53. Which of the following is an unstated assumption made by the author of the passage?
   A. The electron microscope has proven to be an extremely important invention for the scientific community.
   B. The invention of the electron microscope would have been impossible without the prior invention of the television.
   C. The electron microscope cannot function without projection onto a florescent screen.

D. The transmission electron microscope is inferior to the scanning electron microscope.
E. The electron microscope will soon be an outdated technology.

Question 54 refers to the following passage.

Resulting from the amazing success of WAP (Wireless Application Protocol) in smart phones and hand-held devices, wireless technology can have an amazing impact on your day-to-day life. These technologies help to make the mobile information society happen by blurring the boundaries between home, the office, and the outside world. The seamless integration and connectivity that wireless technology brings with it make it possible to work more efficiently. Business users can explore a wide range of interactive services which were difficult to envisage years ago because of the complexity previously involved in making such devices communicate with each other.

54. In order to evaluate the author's claim that wireless technology makes it possible to work more efficiently, it would be most helpful to know which of the following?
A. How often business users access interactive services through their mobile devices
B. What the dollar amount of sales for WAP devices has been in the past few years
C. How wireless technology is used when people are working from home
D. How much less time and effort are required for various tasks at work because of wireless technology
E. What level of operational complexity exists among various electronic devices at the present time

Question 55 refers to the following passage.

A much-loved classic of children's literature, *Alice in Wonderland* portrays a magical world inhabited by the Mad Hatter, the Cheshire Cat, the White Rabbit, the Dormouse, and the Queen of Hearts. Based on the nursery tales that the author Lewis Carroll told to the daughter of one of his academic colleagues, the story is one part of a series that was entitled *Alice's Adventures*. The sequels, *Through the Looking Glass* and *What Alice Did Next* were also extremely popular at the time they were written. The author, whose real name was Charles Dodson, was also a gifted mathematician and enthusiastic amateur photographer.

55. The author includes the final sentence of the passage in order to:
A. point out a lesser-known fact about the subject.
B. state an exception to a general rule.
C. cast doubt on a previous assertion.
D. dispel a commonly-held falsehood.
E. support the veracity of his earlier statements.

Question 56 refers to the following passage.

Do mice really prefer cheese to all other foodstuffs? One well-known American exterminator has revealed his secret to catching these pesky rodents: lemon-flavored candy. It appears that the confection has a double advantage. Its sweet smells attracts the mouse much more strongly than does cheese, and its sticky consistency helps to hold the creature captive for the moment it takes for the trap to release. Through logical analogy, we can therefore conclude that it is fallacious to presume that other groups of animals have preferences for certain food groups. For instance, we cannot readily conclude that all dogs would choose meat or that all cats would select milk as their favorite foodstuffs.

56. Which of the following, if true, would most strongly suggest that the logical analogy mentioned in the passage is incorrect?
A. Mice are attracted more to the texture of the candy than to its smell.
B. Some animals have a very acute sense of smell.
C. Many scientific experiments demonstrate that dogs do not prefer the taste and texture of meat to the taste and texture of other food.
D. Independent observations reveal that mice eat cheese as often as they lemon-flavored candy when both foodstuffs are available to them at the same time.
E. Quite a few cats are allergic to milk and other dairy products.

Praxis Core Reading Practice Test 1 – Answer Key

1. C
2. A
3. C
4. E
5. D
6. B
7. A
8. B
9. D
10. C
11. E
12. C
13. A
14. E
15. A
16. B
17. D
18. E
19. D
20. B
21. C
22. A
23. B
24. D
25. A
26. C
27. E
28. D
29. D
30. E
31. A
32. C
33. C
34. B
35. A

36. B
37. E
38. C
39. A
40. A
41. D
42. B
43. E
44. D
45. D
46. B
47. A
48. C
49. E
50. D
51. E
52. C
53. A
54. D
55. A
56. D

Praxis Core Reading Practice Test 1 – Explanations for the Answers

1. The correct answer is C. The passage states that "historical knowledge [. . .] rests upon judgment and interpretation. [. . .] Facts get their importance from what is made of them in interpretation." Interpretation is based on how something is analyzed, as well as the viewpoint of the interpreter. So, historical knowledge hinges on analyses and opinions.

2. The correct answer is A. The passage tells us that "traditionally, a celebrity would gain the title by his or her work or achievements in a particular field of expertise. [. . .] However, as the twenty-first century progresses, a new celebrity has arrived – the nobody." From these statements we can conclude that the writer is lamenting the fact that celebrities nowadays have no real achievements or expertise.

3. The correct answer is C. "Odious" means hated or detestable; "oligarchy" refers to a dominant group, and "sovereign" means powerful. In addition, "disenfranchise" in the first sentences means to deprive individuals of their rights. So, the author is arguing that men have power over women and that this situation is unfair to women.

4. The correct answer is E. The main idea of the passage is that Madame Curie worked assiduously to discover radium and insisted that the element should only be used for medical purposes. We know this because the first paragraph states that "Marie Curie worked around the clock," while the second paragraph states that "she declared that the radioactive element would be used only to treat disease." The other answers provide specific details from the passage, instead of the main idea. Remember that for main idea questions, you need to avoid answers that are too specific.

5. The correct answer is D. The word "quarry" in this context refers to an object that is sought. For vocabulary questions like this one, you need to search for synonyms in the passage. In this passage, we know that Marie Curry was seeking something because the first paragraph uses the words "discover" and "found."

6. The correct answer is B. Radium is used as a cancer treatment because it destroys cancerous cells. The last sentence explains that radium "eradicates malignant cells without any lasting ill effects on the body." "Eradicate" and "destroy" are synonyms. You may be tempted to choose answer C, but it is incorrect because radium can have positive long-term effects for cancer patients.

7. The correct answer is A. The author notes that "Curie won the Nobel Prize for Chemistry for her discovery" in order to emphasize the importance of her achievement. Prizes like the Nobel Prize are awarded when someone has accomplished something significant or important.

8. The correct answer is B. The background to a discovery is discussed in paragraph 1, and the reasons for its current use are explained in paragraph 2. Paragraph one describes the scientist's work history and professional background. Paragraph 2 describes how the discovery is used today.

9. The correct answer is D. We can conclude that Michelangelo sometimes felt isolated and overwhelmed by the project because the passage states that the artist was "described as the lonely genius" and that he "often felt incompetent to complete the project." In this context, "isolated" means lonely, while being overwhelmed can stem from feelings of incompetence.

10. The correct answer is C. The author mentions Pope Julius II in order to support his assertion about the historical importance of the work of art. The word "entrusted" shows that the Pope was a powerful figure. Since the decision to award the job to Michelangelo was made by a powerful person, we can assume that the work has great historical significance.

11. The correct answer is E. The writer explains that the painting of the ceiling of the Sistine Chapel began in 1508, and then describes the painting as "one of the most beautiful works in art history." Because the writer begins the passage with the date the work began, he is emphasizing that Michelangelo's art is an artistic treasure that has endured for centuries.

12. The correct answer is C. The author mentions that nine scenes were painted on the ceiling of the Sistine Chapel to exemplify the enormity of the project. We know that this was a large-scale undertaking because the work is described as a "massive project" in the first sentence.

13. The correct answer is A. The first paragraph explains: "For extra protection, the tanks are double-walled and surrounded by a concrete covering that is one meter thick." The other answers describe aspects of the storage of the waste, rather than the prevention of leaks.

14. The correct answer is E. "This strategy" in the second paragraph of the passage refers to the process of cooling the liquid by means of cold water. For questions like this one, look carefully at the sentence preceding the phrase. The sentence before the words "this strategy" states that "the liquid needs to be cooled by pumping cold water into coils inside the tanks." So, we know that we are talking about the process of cooling the liquid at this point in the passage.

15. The correct answer is A. The author's attitude can best be described as one of apprehension. "Apprehension" means to be worried about something or to sense that a negative consequence may occur because of something. The author's concern is evident in the first sentence of the passage, which states: "Highly concentrated radioactive waste is lethal and can remain so for thousands of years."

16. The correct answer is B. The primary purpose of the first paragraph is to explain the basic tenets of Marxism, before going on to discuss Marxist views on capitalism and the consequences of private capital. We know this because the second sentence begins with the phrase "according to Marxism." Answer A is too general, and answers C and D are too specific. Answer E is incorrect because the passage talks about Marxism from a political, rather than a historical perspective.

17. The correct answer is D. The first paragraph gives the background to the topic in a general way, and the second paragraph provides specific details about the topic. You may be tempted to choose answer C, but the criticism is only one aspect of the information provided in paragraph 2.

18. The correct answer is E. The writer mentions the "huge impact" that these writings have had on culture and politics in the last sentence in order to juxtapose this impact to Marx's failure to include pragmatic instructions in his work. We know that the author is making a juxtaposition or comparison because the sentence begins with the word "yet."

19. The correct answer is D. The passage provides information for most fully the question: "What is the background to and definition of the term 'liberal arts education'?" The other questions relate only to specific points from the passage, instead of the entire passage.

20. The correct answer is B. The speaker most likely mentions the Magna Carta, the Declaration of the Rights of Man, and the Bill of Rights in order to emphasize the historical importance of this event. We can understand this attitude because the speaker begins her address by stating that "we stand today at the threshold of a great event."

21. The correct answer is C. The word "commodious" means spacious. This is in contrast to the word "small" at the beginning of the sentence.

22. The correct answer is A. The second locomotive that Stephenson invented was an improvement on his first because it ran with greater force and speed. The last sentence of the passage states that "Stephenson's second 'iron horse' was even faster and more powerful than his first creation."

23. The correct answer is B. From the information contained in the passage, it seems reasonable to infer that George Stephenson's inventions laid the basic foundations for modern day public trains and railways. The passage describes how George Stephenson invented the steam locomotive and the world's first public railway. Such inventions lay the basic foundations, which can later be improved upon with advances in technology.

24. The correct answer is D. According to the passage, the metallurgical composition of a coin is determined to be correct by the electricity that has passed through the magnet. The passage states: "Electricity passes through the magnet, causing the coin to slow down in some cases. If the coin begins to slow down, its metallurgic composition has been deemed to be correct." That is to say, the coin slows down because of the electricity that has passed through the magnet.

25. The correct answer is A. It can be inferred that patients today would most likely respond to treatments of the past with fear. We can assume that burning the skin was feared because it is described as a "so-

called" treatment. In addition, the second sentence of paragraph 2 implies that these treatments were inhumane.

26. The correct answer is C. The primary purpose of the passage is to discuss Pasteur's discovery of the rabies vaccine. Paragraph 1 focuses on Pasteur's research on rabies. Paragraph 2 describes how the discovery of the rabies vaccine was made.

27. The correct answer is E. The passage suggests that the discovery of the rabies vaccine was significant because it helped many people avoid physical suffering and death. We know this because paragraph 1 explains that patients with rabies would suffer from "convulsions and delirium, and it would be too late to administer any remedy." The phrase "too late to administer any remedy" indicates that the patient would die from the infection.

28. The correct answer is D. The closest example to the use of a vaccine, as it is described in the passage, is children who get injections to prevent catching mumps and measles. It is the only example from the answer choices involving taking medicine beforehand in order to prevent catching a disease.

29. The correct answer is D. Passage 2 provides an argument that undermines the viewpoint expressed in Passage 1. The first passage argues that people can simply choose to be happy, but the second passage claims that "life is full of stresses and strains that are often not of our own choosing."

30. The correct answer is E. Both authors agree that worry has a direct impact on happiness. The author of passage 1 states that "being happy occurs when we choose not to worry." The author of passage 2 points out that worrying can lead to certain diseases which cause death.

31. The correct answer is A. The use of the phrase "choose to be happy" in both passages serves to convey two differing opinions on whether a happy, worry-free life can really be achieved. Passage 1 promotes happiness as a choice through the rhetorical questions: "So, isn't the choice simple really? Shouldn't we choose to be happy?" Passage 2 refutes this idea by posing another rhetorical question: But isn't McFerrin's refrain that everyone can choose to be happy by simply deciding not to worry overly simplistic in reality?" In addition, we already know that the passages convey two different opinions because of the answer to question 29 above.

32. The correct answer is C. The writer of Passage 1 would most likely respond to this statement by saying that "It may be true that we cannot choose certain events in life, but we can still choose to be happy." We can understand that the writer of passage 1 agrees with Abraham Lincoln's statement that "happiness is a choice for most people."

33. The correct answer is C. Both authors pose rhetorical questions to the reader. We have talked about these rhetorical questions in the answer to question 31 above.

34. The correct answer is B. According to the passage, one way to distinguish a traditional actor from a modern one is by considering the emotional response of the viewers. The passage explains: "Traditionally, actors have provided this emotional release, but this is far from the case in the performances in many modern motion pictures and television programs."

35. The correct answer is A. The speaker's attitude toward the witnesses is best described as one of disrespect. The speaker asserts that "some of the witnesses in this case are not very well and others are awfully ignorant." The speaker therefore disrespects the witnesses by calling them ignorant.

36. The correct answer is B. According to the passage, the best definition of the input hypothesis is that it is a theory which asserts that learners can best improve their language skills when their learning is appropriately challenging. The passage tells us that the learner should be exposed to "lectures or reading materials that are one level above the learner's current level of language ability." If the materials are a level above the learner's current level, they will challenge the learner.

37. The correct answer is E. According to the passage, the two tools which were used to place the stones into their final positions on the pyramid were made from wood. The last two sentences of the passage explain that wooden rods and wooden rockers were used to put the stones in the correct position.

38. The correct answer is C. The writer's main purpose is to give an overview of some of the main implements that were used to construct the Giza pyramids. The passage is devoted to describing the tools that were used during the project.

39. The correct answer is A. The passage suggests that the Giza pyramids are significant because it is incredible that the Egyptians were able to construct the pyramids using only hand-made tools. The passage states that "it is notable that the Egyptians had only the most primitive, handmade tools to complete the massive project." The words "notable" and "massive" support the idea that the construction of the pyramids was incredible.

40. The correct answer is A. In the second paragraph, the author suggests that the effect of social media on relationships has largely been deleterious, which means harmful. The author asserts that "social media platforms may be having a detrimental impact upon our social skills and interpersonal relationships."

41. The correct answer is D. The author would most likely recommend that users of social media turn off their hand-held devices when they are out with friends. The author points out that a "survey revealed that people often remain glued to their hand-held devices to check their social media when they do go out with friends."

42. The correct answer is B. Statistics revealing that most children would be reluctant to visit relatives who do not allow the use of smart phones while at the table would best support the argument presented in the passage. These statistics would support the claim in the passage that "survey results indicate that many people would prefer to interact on Facebook or Twitter, rather than see friends and family in person."

43. The correct answer is E. From the information provided in the chart, we can conclude that Nassir writes his work hours on his calendar only when they vary from 8 am to 5 pm. The facts at the top of the chart tell us that "his normal work schedule every week is from Monday to Friday, 8 am to 5 pm." He has made notes on his calendar for the week beginning on the 9th and the week beginning on the 23rd, so we can surmise that his normal work schedule has been changed for those two weeks.

44. The correct answer is D. From the information provided, we can conclude that Nassir likes to exercise with friends when possible. On the 14th, he is going on a hike with Martin and Suki, and on the 28th he is meeting Jim for a bike ride.

45. The correct answer is D. During the month of February, Nasir will take part in personal training at the gym four times.

46. The correct answer is B. In the past, countries did not have constitutions or other established means to regulate the declaration of martial law. The second paragraph explains that "the constitutions of many countries now make provisions for the introduction of martial law." The use of the word "now" suggests that these provisions were not in place in the past.

47. The correct answer is A. It can be inferred from the passage that the United States Secretary of State would agree with the statement that the declaration of martial law is sometimes needed, although it usually undesirable. We know this because the last sentence of the passage states that, according to the United States Secretary of State, martial law is a "necessary albeit inimical outcome." "Inimical" means undesirable or unfavorable.

48. The correct answer is C. The last sentence of the first paragraph states that a coup is "defined as an illegal and usually violent seizure of a government by a select group of individuals." The use of the word "violent" suggests that other people may be killed or harmed.

49. The correct answer is E. The author finds fault with the civil order control function for its failure to answer the question: "Why do democratic nations sometimes deploy the military to impose order on their populations?" In paragraph 1, the writer explains that "the civil order control function suggests that public order is best maintained through agencies other than the police force or militia." However, the writer explains in paragraph 2 that "in spite of these democratic systems being in place, forms of military control

are still instituted during times of crisis, with a country's military system being mobilized to support civil authorities, such as municipalities and local police forces."

50. The correct answer is D. It can be inferred that the author of the passage would agree with the basic tenet of the civil order control function because there are occasions when public order can only be reinstated through the establishment of military rule, in spite of the disadvantages in doing so. In paragraph 1, the author writes that "martial law, the establishment of military rule over a civilian population, is only imposed when other methods of civil control have proven ineffective." So, the author believes that martial law should only be used when the civil order control function, which is usually the best option, has failed.

51. The correct answer is E. The last step in the process for forming the beam is that the electrons pass through magnetic coils. This answer is found in the following sentence of paragraph 1: "Before traveling through a vacuum in order to remove oxygen molecules, the electrons are focused into a beam by way of magnetic coils." Note that the question focuses on the process of forming the beam in particular, not on the movement of electrons in general, so answer C is incorrect.

52. The correct answer is C. The words "apparatus" and "device" are synonyms in this context. The passage compares the microscope to an old-fashioned television tube in the third sentence of paragraph 1. Both of these items are electronic devices. Machines are larger than devices, so answer A is not the best answer.

53. The correct answer is A. An unstated assumption made by the author is that the electron microscope has proven to be an extremely important invention for the scientific community. This answer is supported by the second sentence of paragraph 1: "This innovation [i.e., the electron microscope] has thereby allowed scientists to study the precise molecules that constitute human life."

54. The correct answer is D. In order to evaluate the author's claim that wireless technology makes it possible to work more efficiently, it would be most helpful to know how much less time and effort are required for various tasks at work because of wireless technology. "Efficient" means to perform with the best use of time and effort.

55. The correct answer is A. The final sentence of the passage is as follows: "The author, whose real name was Charles Dodson, was also a gifted mathematician and enthusiastic amateur photographer." The author includes the final sentence of the passage in order to point out a lesser-known fact about the subject. Many people will know that Lewis Carroll is the author of *Alice in Wonderland.* However, they may not know Lewis Carroll's real name or his other skills.

56. The correct answer is D. The passage states: "Through logical analogy, we can therefore conclude that it is fallacious to presume that other groups of animals have preferences for certain food groups." The logical analogy mentioned in the passage would be incorrect if it were true that mice eat cheese as often as they lemon-flavored candy when both foodstuffs are available to them at the same time. If the mice eat both food groups equally, they would not have a preference, and the logical analogy relies upon the existence of this preference.

Praxis Core Reading Practice Test 2

Questions 1 to 3 refer to the following passage.

Acid has been present in rain for millennia, naturally occurring from volcanoes and plankton. However, scientific research shows that the acid content of rain has increased dramatically over the past two hundred years, in spite of humanity's recent attempts to control the problem. Rain consists of two elements, nitrogen and sulfur. When sulfur is burned, it transforms into sulfur dioxide. Nitrogen also oxides when burned. When released from factories into the atmosphere, both sulfur dioxide and nitrogen oxide react with the water molecules in rain to form sulfuric acid and nitric acid, respectively.

Factories and other enterprises have built high chimneys in an attempt to carry these gases away from urban areas. Nevertheless, the effect of the structures has been to spread the gases more thinly and widely in the atmosphere, thereby exacerbating the problem. The acid in rain also emanates from automobile exhaust, domestic residences, and power stations. *The latter have been the culprit of the bulk of the acid in rainwater* in recent years. Since the pollutants are carried by the wind, countries can experience acid rain from pollution that was generated in countries thousands of miles away.

1. Which one of the following could be substituted for the phrase "the latter have been the culprit of the bulk of the acid in rainwater" in the second to the last sentence with the least change in meaning?
   A. Automobile exhaust has caused the majority of acid rain [...]
   B. Automobile exhaust, domestic residences, and power stations have equally contributed to the creation of acid rain [...]
   C. Power stations have been more widespread geographically than other causes of acid rain [...]
   D. Power stations have generated a great deal of pollution carried by the wind [...]
   E. Power stations have been the largest contributor to the problem [...]

2. Which of the following best describes the organization of the passage?
   A. Scientific explanation and current problems
   B. Chemical analysis and scientific inquiry
   C. Historical background and current problems
   D. Scientific inquiry and possible solutions
   E. Cause and effect

3. Which detail from the passage best supports the primary purpose of the passage?
   A. When sulfur is burned, it transforms into sulfur dioxide.
   B. When released from factories into the atmosphere, both sulfur dioxide and nitrogen oxide react with the water molecules in rain to form sulfuric acid and nitric acid, respectively.
   C. Nevertheless, the effect of the structures has been to spread the gases more thinly and widely in the atmosphere, thereby exacerbating the problem.
   D. The acid in rain also emanates from automobile exhaust, domestic residences, and power stations.
   E. Since the pollutants are carried by the wind, countries can experience acid rain from pollution that was generated in countries thousands of miles away.

Question 4 refers to the following passage.

Ludwig von Beethoven was one of the most influential figures in the development of musical forms during the Classical period. Born in Bonn, Germany, the composer became a professional musician before the age of 12. After studying under both Mozart and Haydn, Beethoven became a virtuoso pianist and had many wealthy patrons, who supported him financially. His most popular works are considered to be his fifth and sixth symphonies, and his only opera is entitled *Fidelio.* It is generally agreed that his compositions express the creative energy of the artist himself, rather than being written to suit the demands of his patrons.

4. The primary purpose of the passage is to:
A. suggest that the works of Beethoven, Mozart, and Haydn are very similar.
B. explore the development of musical composition during the Classical period.
C. provide background information about Beethoven's life and work.
D. explain how Beethoven acquired many wealthy patrons.
E. illustrate the creative energy in Beethoven's symphonies and operas.

Question 5 refers to the following passage.

Painted by the Norwegian artist Edvard Munch, *The Scream* depicts the skeletal face of a person in clear psychological distress. Contrasted against a serene background of asymmetrical red and yellow swirls that represent the sunset, the desperation in the facial characteristics of the subject is said to express humanity's reaction to the anxieties of modern life. Completing the work at the age of 29, Munch admitted that he felt as if a scream went through himself during that time since he was in a state of poor mental and physical health while painting the piece.

5. According to the passage, which one of the following factors most influenced Munch's painting of *The Scream*?
A. his age at the time of working on the painting
B. his own lack of psychological and physiological well-being
C. humanity's experiences of the anxieties of modern life
D. the colors of the sunset
E. the asymmetry of his artistic technique

Questions 6 to 10 refer to the following passage.

Today archeologists are still endeavoring to uncover the secrets of Africa's past. Evidence of the earliest human activity has been found in the south and east of the continent, where climatic conditions helped to preserve the human skeletons and stone tools found there. Genetic science confirms that these are quite likely the oldest remains in the world of modern people, with this classification based on the ability of humans to become adaptable and ready to respond to environmental change. Even though the artifacts and skeletons of early Africans are most commonly found in a highly fragmented state, these findings are more than sufficient in order to make a number of significant conclusions.

Perhaps the most important discovery is that there is great *variation* among the human remains, indicating a wide array of physical differences among members of the population. While the early population was diverse, it has been well established that the earliest species of hominids spread from Africa to other continents. The first traces of human technology, consisting of simple stone tools, were also discovered in Africa. Having been developed long before the invention of metallurgy, tools had gradually become smaller and more sophisticated. Microliths, fine stone tools that were fitted to handles, were used as cutting and scraping tools and may even have been the precursor to the bow and arrow.

6. Which of the following best describes the organization of the passage?
A. A common fallacy is described, and then it is refuted.
B. An unresolved question is posed, and then it is answered.
C. A problem is described, and then a solution is discussed.
D. General information about the research is provided, and then the specific findings of the research are presented.
E. An archeological method is analyzed, and then a contributing phenomenon is discussed.

7. The author states that "genetic science confirms that these are quite likely the oldest remains in the world of modern people" in paragraph 1 primarily in order to emphasize:
A. the depth and breadth of Africa's history.
B. the way that climatic conditions can help to preserve skeletons.
C. the importance of the stone tools found in African sites.
D. the significance of archeological discoveries in Africa.
E. the fact that human beings are adaptable and responsive.

8. As used in paragraph 2, the word "variation" most likely means:
   A. distinction
   B. discrepancy
   C. possibility for error
   D. changeability
   E. variety

9. From the passage, it can be inferred that some of the archeological discoveries from Africa:
   A. were broken into small pieces or extremely damaged.
   B. would not have been located without modern genetic science.
   C. were not as important as those from other continents.
   D. supported the development of metallurgy.
   E. were from inhabitants originally from other continents.

10. The passage suggests that the discovery of microliths was significant for which one of the following reasons?
   A. Microliths illustrate the importance of the invention of the bow and arrow.
   B. Microliths demonstrate the level of sophistication and ingenuity of the prehistoric African population.
   C. Microliths support the view that hominids spread to other continents.
   D. Microliths show that technology at that time consisted of more than stone tools.
   E. Microliths provide evidence for the conclusion that there was a wide array of physical differences among members of the population.

Question 11 refers to the following passage.

A true feat of modern engineering, the Alaska Highway was constructed to link Edmonton in Alberta, Canada, to Fairbanks, Alaska. The first step in completing the mammoth project was to plan the exact route that the road was going to take. Ground and aerial surveys were conducted only slightly in advance of the construction of the road, with the survey teams working just miles ahead of the construction crew in some cases. Apart from the challenges inherent in building a road in such inclement conditions, bridges had to be erected and culverts had to be laid in drainage ditches. Swampland along the route was a further complication, and efforts to avoid the waterlogged ground created many bends in the road.

11. The passage provides information for most fully answering which one of the following questions?
   A. Who were the primary proponents of the construction of the Alaska Highway?
   B. Why did the inclement weather hamper the construction of the Alaska Highway?
   C. How was the Alaska Highway planned and constructed?
   D. How are ground and aerial surveys normally conducted?
   E. Why are bridges and culverts necessary when constructing roadways?

Questions 12 to 15 refer to the following passage.

A complex series of interactive patterns govern nearly everything the human body does. We eat to a rhythm and drink, sleep, and even breathe to separate ones. Research shows that the human body clock is affected by three main rhythmic cycles: the rhythm at which the earth revolves on its axis, the monthly revolution of the moon around the earth, and the annual revolution of the earth around the sun. These rhythms create a sense of time that is both physiological as well as mental. Humans feel hungry about every four hours, sleep about eight hours in every 24-hour period, and dream in cycles of approximately 90 minutes each.

These natural rhythms, sometimes called circadian rhythms, are partially controlled by the hypothalamus in the brain. Circadian rhythms help to explain the "lark vs. owl" hypothesis. Larks are those who quite rightly prefer to rise early in the morning and go to bed early, while owls are those who feel at their best at night and stay up too late. *These cycles* explain the phenomenon of jet lag, when the individual's body clock is out of step with the actual clock time in his or her new location in the world. In humans, births and deaths also follow predictable cycles, with most births and deaths occurring between midnight and 6:00 am.

12. In the first paragraph, the author suggests that our mental and physiological sense of time is:
   A. appropriate
   B. exaggerated
   C. oversimplified
   D. overgeneralized
   E. hypochondriacal

13. In the second to the last sentence of the passage, the phrase "these cycles" refers to:
   A. the "lark vs. owl" hypothesis
   B. circadian rhythms
   C. the hypothalamus in the brain
   D. the individual's body clock
   E. cycles of birth and death

14. The author's attitude toward owls in the "lark vs. owl" hypothesis can best be described as one of:
   A. disapproval
   B. skepticism
   C. hostility
   D. support
   E. fanaticism

15. The author would most likely recommend that sufferers of jet lag do which of the following?
   A. Better control their circadian rhythms.
   B. Take medicine to regulate the hypothalamus.
   C. Go to bed earlier than usual.
   D. Stay up later than usual.
   E. Allow their body clocks to adjust to the time difference naturally.

Question 16 refers to the following passage.

Depicting the events of a single day, James Joyce's epic novel *Ulysses* took more than 20,000 hours, or a total of eight years, to write. Set in Dublin, the novel was initially published in installments as a series before the Parisian publishing house Shakespeare and Company issued a limited edition of 1,000 copies. The book was risqué for its time, and was classified as obscene material in the United Stated. After the work was cleared of obscenity charges, an unexpurgated version was accepted for publication by Random House in New York. Ironically, it was not available in Dublin until 40 years later.

16. The author suggests which of the following about the fact that *Ulysses* was published in Dublin 40 years after it was released in New York?
   A. Irish publishing companies often engage in dilatory practices when dealing with their authors.
   B. Irish publishers were dissuaded in publishing the novel since it depicted the events of only one day.
   C. Random House did not have a division in Dublin at that time.
   D. Social mores in Dublin were much stricter than those of the United States at that time.
   E. Dublin had a more liberal society than that of Paris.

Question 17 refers to the following passage.

Gibberellins are a complex group of plant hormones that are involved in many botanical processes. Commonly used in combination with similar botanical hormones called auxins, their primary function is to promote plant growth by controlling the elongation of cells. They also promote the formation of fruit and seed, as well as delay aging in leaves. Having become important for commercial reasons in recent years, the hormones are also used to help meet the ever-growing demand for new hybrids of plants and flowers.

17. Which of the following best describes the botanical significance of gibberellins?
   A. Without them, plant hormones would be involved in more processes.
   B. Because of gibberellins, plants cells enlarge, thereby causing plants to grow.

C. Leaves age more quickly, owing to the function of gibberellins.
D. Gibberellins have nocuous consequences for fruits and seeds.
E. The demand for new plants and flowers would not have grown so extensively without gibberellins.

Question 18 refers to the following passage.

For every building that is successfully constructed, there are countless others that have never received the chance to leave the drawing board. Some of these unbuilt structures were practical and mundane, while others expressed the flights of fancy of the architect. Known to us today only through the plans left on paper, many unbuilt buildings were originally designed to commemorate particular people or events. Such was the case with the monument dubbed the *Beacon of Progress*, which was to be erected in Chicago to display exhibits dedicated to great Americans in history. However, scholar Samantha Mulholland points out that other proposed projects were far more quixotic, like that of *The Floating Spheres*, described as modules held aloft by hot air to house cities of the future.

18. Samantha Mulholland suggests that which of the following explains why some proposed projects were never constructed?
A. Some projects were never undertaken due to the fact that they did not commemorate any significant event.
B. The plans for some projects had serious design flaws.
C. Some projects were too extravagant and impractical ever to be built.
D. People were not ready to face the future of housing at the time that the construction of *The Floating Spheres* was proposed.
E. *The Floating Spheres* would have been built had it included an important monument.

Question 19 refers to the following passage.

The ancient legal code of Babylonia had severe sanctions for a wide range of crimes. Perhaps best viewed as a way to express personal vengeance, punishments included cutting off the fingers of boys who had hit their fathers or gouging out the eyes of those who had blinded another person. As with most ancient peoples, the Babylonians did not believe in humane treatments for offenders. Sumerian King Ur Nammu, who formulated a set of laws that were surprisingly modern in their approach, did not follow these draconian forms of retribution. Sumerian law stipulated that perpetrators of violent crimes pay monetary damages to their victims, and Ur Nammu's system is the first recorded example of financial awards being imposed in lieu of other forms of punishment.

19. The author mentions Sumerian King Ur Nammu primarily in order to:
A. criticize previous Babylonian rulers.
B. emphasize the severity of the Babylonian system of justice.
C. imply that Babylonian sanctions were just for their time.
D. point out that humane treatments were not really suitable during that epoch.
E. provide a contrast with the forms of punishment meted out by the Babylonians.

Questions 20 to 23 refer to the following passage.

In his book *Il Milione*, known in English as *The Travels of Marco Polo*, the intrepid explorer describes the marvels he encountered as he journeyed to China. Upon his visit to the emperor Kublai Khan in Cathay, Polo witnessed the magical illusions performed by the court wizards of the supreme ruler. Watching in amazement as the wizards recited incantations, Polo first saw a row of golden cups *levitate* over the table as Khan drank from each one without spilling a drop. Polo also recounted that Khan had astonishing powers over wild animals. Unrestrained and ostensibly obedient, lions would appear to lie down in humility in front of the emperor.

However, Khan was venerated for much more than these acts of mere wizardry. Polo's account tells us that the ruler presided over an extremely modern state. Paper currency, integrated with seals of authenticity to prevent counterfeiting, existed during Khan's rule. In addition, his establishment of a vast postal system meant that he would receive news in a fraction of the time that was normally required. Under the rule of Khan, the roads of the empire were also well-maintained, and travelers could reach their destinations relatively quickly and efficiently. Although some academics have disputed the veracity of Polo's written account of the Khan Empire, common sense tells us that there would have been little motive for the explorer to have exaggerated his version of events since he was being held captive at the time with no hope of release.

20. It can be inferred from the passage that the primary reason why the court wizards performed magical illusions was to:
    A. venerate the majesty of Kublai Khan.
    B. play a trick on Marco Polo.
    C. provide an interesting story for the book *Il Milione*.
    D. make Kublai Khan and his court appear powerful and mysterious.
    E. help the ruler control wild animals.

21. The author most probably uses the word "levitate" in paragraph 1 to mean:
    A. rise
    B. drag
    C. hover
    D. linger
    E. hang up

22. Some academics find fault with *Il Milione* for its failure to answer which of the following questions?
    A. Why does Marco Polo's account go against common sense?
    B. Did paper currency really exist during Khan's rule?
    C. Was Khan's state as modern as Polo described?
    D. How could Kahn have established such an extensive postal system?
    E. Why should we believe Polo's version of events?

23. Which of the following best describes the organization of the passage?
    A. It discusses a problem and then provides a solution.
    B. It recounts a story and then offers an explanation.
    C. It describes a social phenomenon and then illustrates it.
    D. It gives the historical background to a piece of writing and then provides further details about it.
    E. It compares one version of a historical event to a differing account and interpretation of the event.

Questions 24 to 26 refer to the following passage.

Two original forms of theater have emerged from Japanese culture: Noh and Kabuki. Noh, the older form, was originally established to meet the demands of the "discriminating Japanese aristocracy" and remained "unchanged for more than six centuries." Noh renders mundane, everyday activities, like drinking tea or arranging flowers, into exquisite artistic performances. Consisting of minimal spectacle, bare stage designs, and little spoken dialogue, Noh is classified as more ritual than drama. In order to convey the dialogue, a chorus sings the protagonist's lines while the performer engages in the "solemn act" of the dance.

Kabuki performances are discernably different than those of Noh. Based on puppet theater, Kabuki is designed to meet the tastes of the general populace, rather than those of the aristocracy. According to long-standing theatrical custom, Kabuki performances can be extremely long, lasting up to twelve hours in some cases. Since movement plays a greater role than dialogue, Kabuki actors must wear heavy makeup and engage in highly stylized actions. Because of its appeal to the general populace, Kabuki theater remains as fascinating and exotic as it has always been, even though its purity has been somewhat compromised through exposure to other cultures.

24. The use of quotations in the passage suggests which of the following about followers of Noh?
A. They lament the fact that Noh clings on to outdated customs of the past.
B. They believe that Kabuki theater is overtly flamboyant.
C. They fear that the popularity of Kabuki theater may diminish the appeal of Noh.
D. They plan to make Noh more up-to-date in order to increase its following.
E. They want to emphasize that followers of Noh are traditional, discerning, and serious.

25. The second paragraph implies that Japanese audiences today would respond to Kabuki theater with:
A. admiration
B. impatience
C. confusion
D. boredom
E. detachment

26. Followers of Noh and followers of Kabuki would probably agree with which one of the following statements?
A. Theatrical productions sometimes last too long.
B. Japanese theater is unlikely to change in the future.
C. Theatrical performances should be highly stylized and full of spectacle in order to be effective.
D. Japanese theater is an important and interesting aspect of Japanese culture.
E. The purity of all forms of Japanese theater has been compromised through exposure to other cultures.

Questions 27 to 29 refer to the following passage.

In Southern Spain and France, Stone Age artists painted stunning drawings on the walls of caves nearly 30,000 years ago. Painting pictures of the animals upon which they relied for food, the artists worked by the faint light of lamps that were made of animal fat and twigs.
In addition to having to work in relative darkness, the artists had to endure great physical discomfort since the inner chambers of the caves were sometimes less than one meter in height. Thus, the artists were required to crouch or squat uncomfortably as they practiced their craft.

Their paints were mixed from natural elements such as yellow ochre, clay, calcium carbonate, and iron oxide. However, many other natural elements and minerals were not used. An analysis of the cave paintings reveals that the colors of the paints used by the artists ranged from light yellow to dark black. The artists utilized ochre and manganese as engraving tools in order first to etch their outlines on the walls of the caves. Before removing their lamps and leaving their creations to dry, they painted the walls with brushes of animal hair or feathers. Archeologists have also discovered that ladders and scaffolding were used in higher areas of the caves.

27. What was the last step in the process of Stone Age cave drawings?
A. The paintings were etched.
B. The paint was applied.
C. The lamps were removed.
D. The artwork was left to dry.
E. The scaffolding was erected.

28. Which of the following best expresses the attitude of the writer?
A. It is surprising that the tools of Stone Age artists were similar to those that artists use today.
B. It is amazing that Stone Age artists were able to paint such beautiful creations in spite of the extreme conditions they faced.
C. The lack of light in the caves had an effect on the esthetic quality of the paintings.

D. It is predictable and banal that Stone Age artists would paint pictures of animals.
E. The use of natural elements in paint was not an environmentally-friendly practice.

29. Which of the following best supports the main idea of paragraph 2?
A. Thus, the artists were required to crouch or squat uncomfortably as they practiced their craft.
B. Their paints were mixed from natural elements such as yellow ochre, clay, calcium carbonate, and iron oxide.
C. However, many other natural elements and minerals were not used.
D. An analysis of the cave paintings reveals that the colors of the paints used by the artists ranged from light yellow to dark black.
E. The artists utilized ochre and manganese as engraving tools in order first to etch their outlines on the walls of the caves.

Questions 30 to 35 refer to the following pair of passages.

Passage 1
Credit card debt is a major cause of over one million bankruptcies each year. The reason is that many people get a credit card on impulse and fail to read the terms and conditions. By the time annual fees are accrued, payments can be missed, which causes balances to skyrocket.
Although we all would like to believe that credit card companies are culpable, individuals themselves are the real *culprits*. In short, if your *credit card debt* is out of control, the real cause of your financial mess is you. If you can summon enough willpower and strength to manage your finances and spending, then you will find yourself the winner in the game of finance. It may be easy to get into debt, but getting out of debt is much more difficult. One simple phrase sums up the solution to financial problems: If you don't have the money to spend, then don't spend it.

Passage 2
It has to be said that external forces and market conditions have a huge impact on personal financial situations. Have you ever noticed that the things you buy at the store go up a few pennies between shopping trips? Not every week and not by much – just little by little – but they continue to creep up. There is a way that the effect of price increases upon personal finances can be minimized: buy in quantity when prices are low. My philosophy is to set out to find the best prices I can get on quantity purchases of things such as bathroom items and dry and canned food, even if I have to increase my *credit card debt* to get them. You will be surprised by how much you can save, for example, by buying a twenty pound bag of rice as opposed to a one pound bag.

30. Which best describes the relationship between Passage 1 and Passage 2?
A. Passage 1 introduces a theory that is discussed at length in Passage 2.
B. Passage 1 recounts a personal experience that is supported by Passage 2.
C. Passage 1 reveals a particular opinion on a topic, while Passage 2 provides an alternative point of view.
D. Passage 2 refutes the scientific evidence that is provided in Passage 1.
E. Passage 2 explicates the hypothesis that is mentioned in Passage 1.

31. The reference to "credit card debt" in the passages serves to:
A. defer to authority.
B. provide a contrast.
C. offer an explanation.
D. reveal a cause.
E. propose a solution.

32. The writer of passage 1 would disagree most strongly with which of the following statements from passage 2?
A. External forces and market conditions have a huge impact on personal financial situations.
B. The things you buy at the store go up a few pennies between shopping trips.

C. There is a way that the effect of price increases upon personal finances can be minimized: buy in quantity when prices are low.
D. My philosophy is to set out to find the best prices I can get on quantity purchases of things such as bathroom items and dry and canned food, even if I have to increase my credit card debt to get them.
E. You will be surprised by how much you can save, for example, by buying a twenty-pound bag of rice as opposed to a one-pound bag.

33. Which of the following words best characterizes how both authors view avoiding unnecessary expenses?
A. advantageous
B. undisciplined
C. victorious
D. philosophical
E. obsequious

34. Both authors do which of the following?
A. Speculate on possible reasons for a phenomenon
B. Refer to personal experiences to support their argument
C. Mention buying in bulk as a means to saving money
D. Provide specific examples to refute the opposing point of view
E. Pose rhetorical questions to the reader

35. In the context of Passage 1, "culprits" most nearly means:
A. criminals
B. corruptors
C. thieves
D. con artists
E. guilty parties

Question 36 refers to the following passage.

Educational psychology studies pupils in a classroom setting in order to help educators to understand the behaviors and attitudes that affect learning and teaching. This branch of psychology was a reaction against the psychometric movement, which tested students in order to place them into "streamed" classes of different ability levels. The popularity of IQ testing and streamed education declined in the second half of the twentieth century, and the education profession is now focused on developing programs that view students as individuals and advising schools how better to function as organizations.

36. According to the passage, the best way to distinguish between the education profession before the second half of the twentieth century and current educational practice would be by:
A. looking at the results of psychometric testing.
B. studying pupils in a classroom setting.
C. supporting the benefits of IQ testing.
D. determining whether students are grouped into categories based on test results.
E. criticizing how the school functions as an entity.

Question 37 refers to the following passage.

Born Marguerite Johnson, Maya Angelou was an American writer, poet, educator, and actor. Her seven autobiographies, based on her experiences of childhood and as a young adult, have been published around the world and have received international acclaim. From her writings, we know that Angelou was abused as a child and had a baby at sixteen, before going on to work as an actress and school administrator. Her writings are widely regarded as celebrating the African-American experience, as well as the capacity not only to survive hardship, but also to thrive and flourish in the face of adversity.

37. The author mentions Angelou's childhood abuse most likely in order to:
A. give an example of the hardship that she faced.
B. provide a contrast to her work as a school administrator.
C. reveal the main basis of her autobiographical writing.
D. suggest the reason why she did not write under her real name.
E. indicate future directions for African-American literature.

Questions 38 to 40 refer to the following passage.

The ancient Egyptians used eye shadow over 5,000 years ago. The cosmetic was used for personal beautification, as well as for practical reasons. Consisting of a paste made from malachite, a copper salt that was bright green, the eye paint protected against glare from the sun, in addition to being an attractive color. On her upper eye lids, Cleopatra wore blue eye shadow made of ground lapis lazuli stone, much like other women of her day.

The queen used green malachite as an accent below her eyes, and kohl, which consisted of lead sulfide, to provide color to her eyelashes and eyebrows. Red ochre, iron-based clay, provided her with lip and cheek color. Henna, a reddish-brown dye that was derived from a bush, was also commonly used by women in those days as a nail polish. The henna was thickened with tannin from the bark or fruit of various trees in order to be suitable for cosmetic use. The use of this particular cosmetic was not limited to women. Men also used the substance to darken their hair and beards.

38. The author would most likely agree with which of the following statements about Cleopatra?
A. Cleopatra was a typical ancient Egyptian woman in many ways.
B. Cleopatra was a trendsetter in beauty and fashion.
C. Cleopatra wore too much make-up for a woman of her social standing.
D. Cleopatra's use of cosmetics reflected the fashion of the times in which she lived.
E. Cleopatra's eye shadow demonstrated that she was a member of a privileged social class.

39. Which of the following best describes the relationship between the two paragraphs in the passage?
A. The first paragraph states the facts, and the second paragraph exemplifies those facts.
B. The first paragraph describes a problem, and the second paragraph explains possible solutions.
C. The first paragraph mentions a historical event, and the second paragraph interprets the event.
D. The first paragraph analyzes a phenomenon, and the second paragraph defends the cultural importance of the phenomenon.
E. The first paragraph provides one example, while the second paragraph gives further examples.

40. What word best describes the style of writing in this passage?
A. argumentative
B. persuasive
C. informative
D. condemning
E. unbalanced

Questions 41 to 45 refer to the following passage.

Dance notation is to choreography what written scores are to music and what written scripts are to drama. The representation of movement in these notation systems varies, although most are based on drawings, stick figures, abbreviations, musical notes, or abstract symbols. Recording the movements of dance through a shortened series of characters or symbols, more than one hundred systems of dance notation have been created over the past few centuries.

In the seventeenth century, Pierre Beauchamp devised a notation system for Baroque dance. Known as Beauchamp-Feuillet notation, his system was used to record dances until the end of the eighteenth century. Later, Vladimir Ivanovich Stepanov, a Russian, was responsible for notating choreographic scores for the famous *Sergevev Ballet Collection*, including works such as *Swan Lake*, *Sleeping Beauty*, and *The Nutcracker*. Thanks to Stepanov's system, dance companies were enabled to stage these works outside of Russia. Hanva Holm was the first choreographer to copyright the notations of her dance scores, securing the rights for *Kiss Me Kate* in 1948. Two other notation systems, Labanotation and Benesh notation, also known as choreology, are in wide-spread use today. Apple created the first computerized system to display an animated figure on the screen that illustrated dance moves, and many other software systems have been developed to facilitate computerized dance notation.

41. The passage is primarily concerned with:
    A. describing the history of dance notation and its use.
    B. illustrating the way in which dance notation has improved performance.
    C. defending changes to various dance notation systems.
    D. criticizing outdated forms of dance notation.
    E. suggesting that choreology is the most popular form of dance notation.

42. The function of the first sentence of the passage (Dance notation is . . . drama.) is to:
    A. provide an example.
    B. set up relevant analogies.
    C. pose a rhetorical question.
    D. state a hyperbole.
    E. compare historical periods.

43. According to the passage, Beauchamp-Feuillet notation differs from Vladimir Ivanovich Stepanov's notation system in that Stepanov's system:
    A. was used for a different genre of dance during a different time period.
    B. was used for works performed in the Russian language.
    C. was never copyrighted.
    D. was also known as choreology.
    E. did not use animation.

44. The passage indicates which of the following about Hanva Holm?
    A. She was a Broadway dancer in her day.
    B. She performed as a dancer in *Kiss Me Kate* in 1948.
    C. She was the first person to register intellectual property rights for a dance notation system.
    D. Her system is still in widespread use today.
    E. Her system was superseded by the Labanotation and Benesh notation systems.

45. The author most likely mentions Apple and other computerized dance notation systems in the last sentence of the passage in order to:
    A. point out the similarities among various computerized systems.
    B. advocate the use of computer software for choreographed performances.
    C. identify the systems that have replaced choreology.
    D. imply that computerized dance notation systems are of better quality than those of the past.
    E. indicate possible trends in dance notation.

Question 46 refers to the following passage.

Although the foundations of the movement can be traced back to the artists van Gogh and Gauguin in the late nineteenth century, the first recorded use of the term Expressionism was in Germany in the early twentieth century. Influencing art, literature, theater, and architecture, Expressionism strives to illustrate the inner emotional reaction to a reality. In this approach, the traditional notion of realism is to be disregarded, as are the conventional ideas of beauty and proportion. Accordingly, Expressionist artists use distortion, incongruous color schemes, and exaggerated shapes and sizes to reveal their emotions.

The impact of the movement is also present in fictional and poetic works of the era, particularly those which represent the dislocation of the individual within society.

46. The passage suggests that Expressionism illustrates which one of the following phenomena?
   A. Artistic movements are ever-changing with the passage of time.
   B. Abstract art is more popular than realistic art.
   C. Human beings felt out of sync with their communities at the time this movement was taking place.
   D. Most twentieth century artists were nonconventional.
   E. Even incongruous images have a certain artistic beauty.

Questions 47 and 48 refer to the following passage.

Known as the Centennial State, Colorado is divided into sixty-three counties. The eastern part of the state was gained by the U.S. in 1803 as part of the Louisiana Purchase, while the western part was acquired from Mexico by treaty in 1848. Colorado joined the union as the 38th state in 1876, shortly after the first substantial discovery of gold in the state near Pikes Peak in 1859. The Rocky Mountains run along a north-south line through the center of the state, and there are several famous national parks and monuments, including Rocky Mountain National Park, Black Canyon of the Gunnison National Park, Mesa Verde, Dinosaur National Monument, and the Great Sand Dunes National Monument.

Agriculture in the state involves the production of wheat, hay, corn, sugar beets, and other crops, as well as cattle ranching and raising other livestock. The packaging, processing, fabrication, and defense industries form the *lion's share of* revenues from business and commerce in the state. Perhaps lesser-known is the fact that Colorado contains the world's largest deposits of molybdenum, a brittle silver-grey metallic chemical element that is used in some alloy steels.

47. Which of the following could be substituted for the phrase "lion's share of" in the second paragraph with the least change in meaning?
   A. preponderance of
   B. superfluous
   C. excessive
   D. abundant
   E. ever-increasing

48. The primary purpose of the passage is to:
   A. discuss trade and commerce in a particular state.
   B. sum up the historical background and notable features of a particular state.
   C. provide pertinent political details about the acquisition of a particular state.
   D. emphasize the importance of agriculture for trade and commerce in Colorado.
   E. reveal a surprise fact about a particular state.

Question 49 refers to the following passage.

Reconstruction is the process whereby words are constructed in an undocumented language by comparing its sound system to that of known related languages. The practice, which is also called internal reconstruction, is based on the postulation that certain sounds have variants in various languages. For instance, the Latin word *pater* and the Gothic word *fadar* show a systematic correspondence between the *p* and *f* sounds in these languages. This leads to the hypothesis that *p* was the earlier variant of the *f* consonant in other related languages, as well as in antediluvian languages and Indo-European forms.

49. Which of the following, if true, would most strengthen the argument presented in the passage?
   A. The *f* sound was an early variant of the *p* sound in antediluvian languages.
   B. Several languages around the world today remain undocumented.
   C. The *c* and *k* sounds are easily distinguished from each other in many languages.
   D. A systematic correspondence exists between the *b* and *v* sounds in certain languages.
   E. Internal reconstruction is useful in many other academic disciplines.

Question 50 refers to the following passage.

The Higgs mechanism is the process in quantum field theory whereby symmetry is broken down, leading to massive particles. Quantum field theory alone tells us that all particles should be massless. Yet, groundbreaking scientific research has found that particles can acquire mass when the symmetry of energy within a system is less than that of the interaction governing the system. Theoretically, scientists understand that the Higgs particle is a by-product of the acquisition of mass by other particles. Discovering this elusive particle remains one of the greatest challenges of modern-day particle physicists.

50. The author most likely elaborates on quantum field theory in the second sentence in order to:
   A. change the subject from mechanisms to particles.
   B. reveal that the Higgs mechanism inheres in a basic contradiction.
   C. illustrate how the symmetry of energy within a system can be lower than that of the governing system.
   D. explain the process by which massive particles are formed.
   E. refute the notion that particles can be by-products of mass.

Questions 51 to 53 refer to the following passage.

The Earth's only natural satellite, the Moon lacks its own atmosphere and is only about one-fourth the size of the planet it orbits. The equality of its orbital rate to that of the Earth is the result of gravitational locking, also known as synchronous rotation. Thus, the same hemisphere of the Moon always faces the earth. The brightest lunar surface areas are formed from meteoric material, while its dark surface regions, called mare basalts or basaltic plains, are the result of volcanic flooding related to impacts from asteroids. Scientific dating of samples from the Moon's crust reveals that the materials range in age from three to four billion years old.

Lunar evolution models suggest that the development of the Moon occurred in five principal stages: (1) increase in mass followed by large-scale melting; (2) separation of the crust with concurrent bombardment by meteors; (3) melting at greater depth; (4) lessening of meteoric bombardment with further melting at depth and the formation of basaltic plains; and (5) the cessation of volcanic activity followed by gradual internal cooling. Because of the geological and mineral composition of the surface of the Moon, one popular theory hypothesizes that the Moon grew out of debris that was dislodged from the Earth's crust following the impact of a large object with the planet.

51. For which of the following situations does the concept of synchronous rotation, as it is defined in the passage, provide the most likely explanation?
   A. The Moon goes through four phases every twenty-eight days.
   B. A star appears to shine at the same intensity, regardless of its position in the sky.
   C. Two objects fall to the ground at the same speed and land at the same time.
   D. An undiscovered planet has two equal hemispheres.
   E. A telecommunications satellite is always in the same position above a certain city on Earth.

52. The passage suggests that which one of the following probably occurred after the completion of the process of lunar evolution?
   A. Ice continued to melt on the surface of the Moon.
   B. The likelihood of the collision of the Moon with a meteor was substantially reduced.
   C. Melting at depth still occurred.
   D. There were further eruptions of magma or lava.
   E The temperature of the internal core of the Moon was lower than it was previously.

53. Which of the following, if true, would tend to disprove the hypothesis that the Moon grew out of debris that was dislodged from the Earth's crust?
   A. An analysis reveals that there are no geological similarities between samples of material from the surface of the Moon and material from the Earth's crust.
   B. The Moon has been found not to have had any previous volcanic activity.

C. Many meteors bombarded with the Earth during the process of lunar evolution.
D. A great deal of debris is created when a meteor collides with the Earth.
E. A sample from the Moon's crust is determined to be three and a half billion years old.

Questions 54 to 56 refer to the following information.

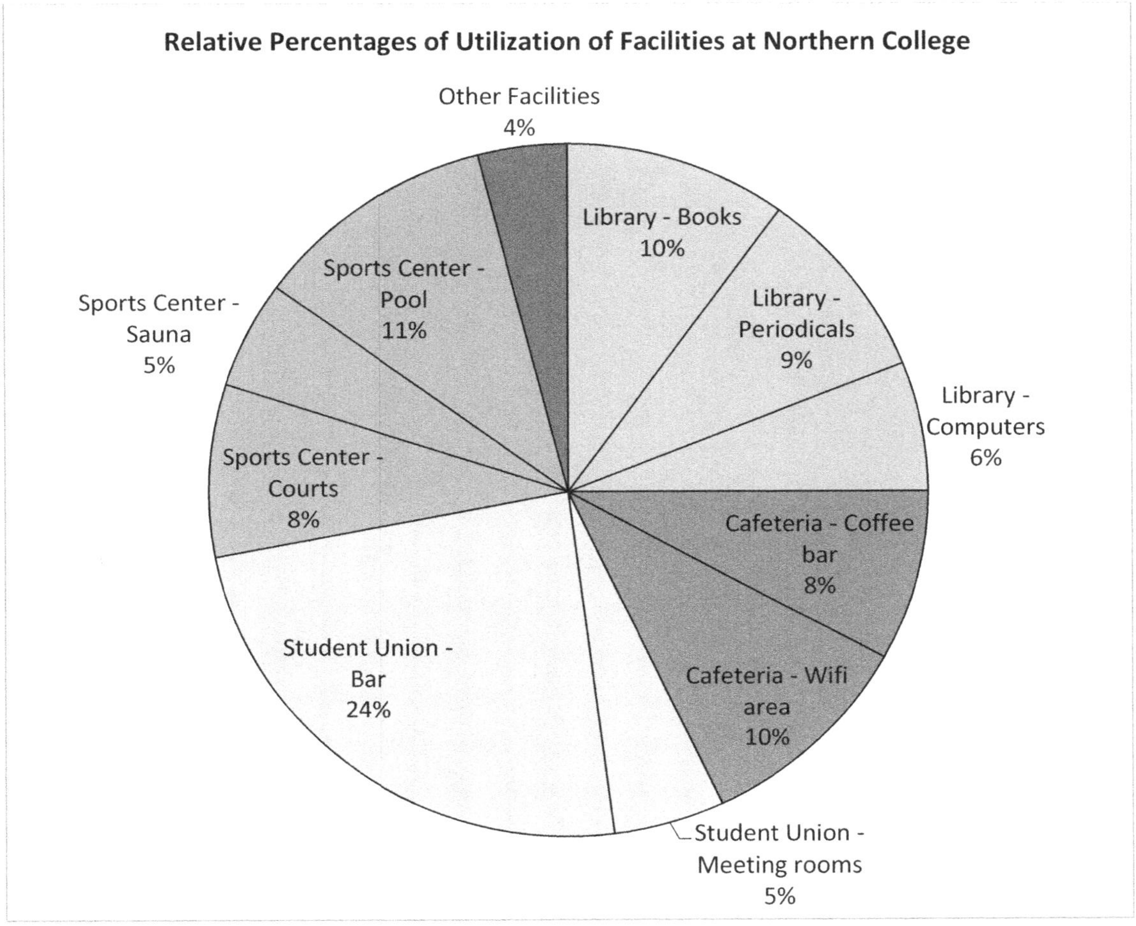

54. Which of the following statements is best supported by the information provided above?
A. More people borrow periodicals from the library than books.
B. The library needs to acquire new computers to meet student demand.
C. The sports center does not have any fitness classes.
D. The wifi area is more popular than the coffee bar.
E. Other facilities suffer from poor resources and underfunding.

55. Which of the following groups, when combined, have a utilization rate of more than 15%? You may select more than one answer.
A. Cafeteria Wifi Area and Other Facilities
B. Student Union Bar and Meeting Rooms
C. Library Periodicals and Computers
D. Sports Center Sauna and Courts
E. Sports Center Sauna and Pool

56. Which one of the facilities has the greatest total utilization rate at Northern College?
A. Library
B. Cafeteria
C. Student Union
D. Sports Center
E. Other

Praxis Core Reading Practice Test 2 – Answer Key

1. E
2. A
3. C
4. C
5. B
6. D
7. D
8. E
9. A
10. B
11. C
12. A
13. B
14. A
15. E
16. D
17. B
18. C
19. E
20. D
21. C
22. E
23. D
24. E
25. A
26. D
27. D
28. B
29. B
30. C
31. B
32. D
33. A
34. A
35. E

36. D
37. A
38. D
39. E
40. C
41. A
42. B
43. A
44. C
45. E
46. C
47. A
48. B
49. D
50. B
51. E
52. E
53. A
54. D
55. B and E
56. C

Praxis Core Reading Practice Test 2 – Explanations for the Answers

1. The correct answer is E. The last paragraph mentions that "the acid in rain also emanates from automobile exhaust, domestic residences, and power stations. The latter have been the culprit of the bulk of the acid in rainwater in recent years." The phrase "the latter" refers to the last thing mentioned, so it refers to power stations at the end of the previous sentence. Therefore, power stations have been the largest contributor to the problem.

2. The correct answer is A. The mention of the chemicals nitrogen and sulfur in paragraph 1 shows that a scientific explanation is being provided. Paragraph 2 talks about "exacerbating the problem," indicating that current problems are being discussed. Accordingly, the organization of the passage is scientific explanation and current problems.

3. The correct answer is C. The statement that best supports the writer's main idea is that "the effect of the structures has been to spread the gases more thinly and widely in the atmosphere, thereby exacerbating the problem." This statement links back to the main idea of the passage, which is stated in paragraph 1, sentence 2: "scientific research shows that the acid content of rain has increased dramatically over the past two hundred years, in spite of humanity's recent attempts to control the problem."

4. The correct answer is C. The primary purpose of the passage is to provide background information about Beethoven's life and work. The passage begins by providing information about the composer's musical training, before going on to talk about his professional life and compositions.

5. The correct answer is B. Munch's own lack of psychological and physiological well-being most influenced his painting of *The Scream*. The last sentence of the passage explains that "Munch admitted that he felt as if a scream went through himself during that time since he was in a state of poor mental and physical health while painting the piece." Note that "physiological" and "physical" are synonyms.

6. The correct answer is D. General information about the research is provided, and then the specific findings of the research are presented. The first paragraph describes the background to archeological research in Africa, and the second paragraph gives specific details about the remains and artifacts that were discovered there.

7. The correct answer is D. The author states that "genetic science confirms that these are quite likely the oldest remains in the world of modern people" in paragraph 1 primarily in order to emphasize the significance of archeological discoveries in Africa. We know this because the paragraph goes on to explain that "these findings are more than sufficient in order to make a number of significant conclusions."

8. The correct answer is E. As used in paragraph 2, the word "variation" most likely means variety. "Variation" is synonymous with the words "wide array" and "diverse," which are used later in paragraph 2.

9. The correct answer is A. From the passage, it can be inferred that some of the archeological discoveries from Africa were broken into small pieces or extremely damaged. The last sentence of paragraph 1 of the passage tells us that "the artifacts and skeletons of early Africans are most commonly found in a highly fragmented state." "Fragmented" means broken into pieces.

10. The correct answer is B. The passage suggests that the discovery of microliths was significant because these tools demonstrate the level of sophistication and ingenuity of the prehistoric African population. The second to last sentence of paragraph 2 states: "Having been developed long before the invention of metallurgy, tools had gradually become smaller and more sophisticated." Microliths are then given as an example of more sophisticated tools.

11. The correct answer is C. The passage provides information for most fully answering the following question: How was the Alaska Highway planned and constructed? The passage explains how the land was surveyed and how the road and other structures were built.

12. The correct answer is A. In the first paragraph, the author suggests that our mental and physiological sense of time is appropriate. The author explains that all human beings have this sense of time. The

author does not criticize this behavior, but rather, provides factual information about the topic. From the tone of the passage, we can therefore surmise that that author views this behavior as appropriate.

13. The correct answer is B. In the second to the last sentence of the passage, the phrase “these cycles” refers to circadian rhythms. The two previous sentences state: “Circadian rhythms help to explain the “lark vs. owl” hypothesis. Larks are those who quite rightly prefer to rise early in the morning and go to bed early, while owls are those who feel at their best at night and stay up too late.” Larks and owls are given as an example, so the phrase “these rhythms” refers back to the subject of circadian rhythms in the previous sentence.

14. The correct answer is A. The author’s attitude toward owls in the “lark vs. owl” hypothesis can best be described as one of disapproval. The author says that larks “quite rightly prefer to rise early in the morning,” but owls “stay up too late.” So, the author disapproves of the owl’s behavior.

15. The correct answer is E. The author would most likely recommend that sufferers of jet lag allow their body clocks to adjust to the time difference naturally. The author begins the second paragraph by explaining that "these natural rhythms, sometimes called circadian rhythms, are partially controlled by the hypothalamus in the brain.” Since the author refers to the rhythms as a natural phenomenon, he or she would most likely suggest that the time difference be overcome naturally.

16. The correct answer is D. The author suggests that social mores in Dublin were much stricter than those of the United States at the time that *Ulysses* was published in New York. The passage tells us that “the book was risqué for its time” and was originally classified as “obscene material.” In this context, the word “mores” means moral views, and the word “risqué” means indecent.

17. The correct answer is B. Gibberellins are of botanical significance because they cause plants cells to enlarge, thereby causing plants to grow. The passage states that the primary function of gibberellins “is to promote plant growth by controlling the elongation of cells.”

18. The correct answer is C. Samantha Mulholland suggests some proposed projects were never constructed because they were too extravagant and impractical ever to be built. The passage states: “Scholar Samantha Mulholland points out other proposed projects were far more quixotic.” The word “quixotic” means extravagant and impractical.

19. The correct answer is E. The author mentions Sumerian King Ur Nammu primarily in order to provide a contrast with the usual forms of punishment meted out by the Babylonians. The passage states that “the Babylonians did not believe in humane treatments for offenders.” However, King Ur Nammu “did not follow these draconian forms of retribution.”

20. The correct answer is D. It can be inferred from the passage that the primary reason why the court wizards performed magical illusions was to make Kublai Khan and his court appear powerful and mysterious. The first paragraph uses the words “amazement” and “astonishing” to express the mysteriousness of the court.

21. The correct answer is C. The author most probably uses the word “levitate” in paragraph 1 to mean hover. The words “levitate” and “hover” both mean to be suspended in midair.

22. The correct answer is E. Some academics find fault with *Il Milione* for its failure to answer the following question: Why should we believe Polo’s version of events? The passage explains that “Although some academics have disputed the veracity of Polo’s written account of the Khan Empire, common sense tells us that there would have been little motive for the explorer to have exaggerated his version of events.” The phrase “dispute the veracity” means that they doubt whether the story is true.

23. The correct answer is D. The passage gives the historical background to a piece of writing and then provides further details about it. Paragraph 1 describes the book *Il Milione*, and paragraph 2 provides some additional information about Polo’s written account of events.

24. The correct answer is E. The use of quotations in the passage suggests that the followers of Noh are traditional, discerning, and serious. Paragraph 1 uses quotation marks when it states that Noh is for the

"discriminating Japanese aristocracy" and that it depicts a "solemn act." The word "aristocracy" indicates that the dance is traditional in nature. "Discriminating" means "discerning," and "solemn" means "serious."

25. The correct answer is A. The second paragraph implies that Japanese audiences today would respond to Kabuki theater with admiration. The last sentence of the second paragraph states: "Because of its appeal to the general populace, Kabuki theater remains as fascinating and exotic as it has always been." We can surmise that people probably admire something that fascinates them.

26. The correct answer is D. Followers of Noh and followers of Kabuki would probably agree that Japanese theater is an important and interesting aspect of Japanese culture. The first sentence of the passage explains that these forms of theater "have emerged from Japanese culture." Since an article has been devoted to this topic, we can assume that followers consider the topic to be an important and interesting aspect of the Japanese culture.

27. The correct answer is D. We need to have a look at the first and second sentences of the last paragraph, which state: "The artists utilized ochre and manganese as engraving tools in order first to etch their outlines on the walls of the caves. Before removing their lamps and leaving their creations to dry, they painted the walls with brushes of animal hair or feathers." Be sure to read sentences like these very carefully. The etching is the first step. The application of the paint is the second step. Removing the lamps is the third step, while leaving the paint to dry is the final step.

28. The correct answer is B. The attitude of the writer is that it is amazing that Stone Age artists were able to paint such beautiful creations in spite of the extreme conditions they faced. For questions like this one, look for adjectives in the passage that give hints about the author's point of view. The phrase "stunning drawings" in paragraph 1 indicates the author's amazement.

29. The correct answer is B. Paragraph 2 focuses on the natural elements that were used in the process of creating the drawings. Therefore, this sentence is most relevant to the main idea of the passage: "Their paints were mixed from natural elements such as yellow ochre, clay, calcium carbonate, and iron oxide."

30. The correct answer is C. Passage 1 reveals a particular opinion on a topic, while passage 2 provides an alternative point of view. Passage 1 is against credit card debt, but passage 2 states that it is acceptable to increase credit card debt in certain circumstances.

31. The correct answer is B. The reference to "credit card debt" in both passages serves to provide a contrast. Passage 1 states that "if your *credit card debt* is out of control, the real cause of your financial mess is you." Passage 2 promotes the philosophy "to set out to find the best prices [. . .] on quantity purchases [. . .] even if I have to increase my *credit card debt* to get them." This again demonstrates that passage 1 is opposes credit card debt, but passage 2 supports it in some cases.

32. The correct answer is D. The writer of selection 1 would disagree most strongly with this statement because it directly supports increasing credit card debt.

33. The correct answer is A. Both authors view avoiding unnecessary expenses as advantageous. The writer of passage 1 supports avoiding unnecessary expenses by controlling spending. The writer of passage 2 states that unnecessary spending can be avoided by making purchases when prices are low.

34. The correct answer is A. Both authors speculate on possible reasons for a phenomenon. The phenomenon that both authors address is the experience of problems with personal finance. The author of passage 1 asserts that individuals themselves are responsible for the phenomenon, while the author of passage 2 believes that "external forces and market conditions have a huge impact on personal financial situations."

35. The correct answer is E. In the context of Passage 1, "culprits" most nearly means guilty parties. Passage 1 asserts: "Although we all would like to believe that credit card companies are culpable, individuals themselves are the real *culprits*." In other words, individuals are responsible for their spending, so they are guilty when they cannot control their spending.

36. The correct answer is D. According to the passage, the best way to distinguish between the education profession before the second half of the twentieth century and current educational practice is by

determining whether students are grouped into categories based on test results. The passage explains that students used to be placed "into 'streamed' classes of different ability levels." However, current educational practice is to "view students as individuals."

37. The correct answer is A. The author mentions Angelou's childhood abuse most likely in order to give an example of the hardship that she faced. After mentioning the examples from Angelou's life, the passage explains that "Her writings are widely regarded as celebrating the African-American experience, as well as the capacity [. . .] to survive hardship."

38. The correct answer is D. The author would most likely agree with the statement that Cleopatra's use of cosmetics reflected the fashion of the times in which she lived. The last sentence of paragraph 1 mentions that "Cleopatra wore blue eye shadow made of ground lapis lazuli stone, much like other women of her day."

39. The correct answer is E. The first paragraph provides one example, while the second paragraph gives further examples. The entire passage talks about cosmetics. Paragraph 1 talks about eye shadow, and paragraph 2 describes other cosmetics, such as eye liner, lip and cheek color, and nail polish.

40. The correct answer is C. The passage is informative because it merely provides information on cosmetics in ancient times, without taking a particular position on the topic.

41. The correct answer is A. The passage is primarily concerned with describing the history of dance notation and its use. The theme of the passage is introduced in the last sentence of the first paragraph, which says that "more than one hundred systems of dance notation have been created over the past few centuries." The phrase "past few centuries" indicates that a historical account is going to be given.

42. The correct answer is B. The function of the first sentence of the passage is to set up relevant analogies. An analogy is a comparison. The passage compares dance notation to written scores and written scripts.

43. The correct answer is A. According to the passage, Beauchamp-Feuillet notation differs from Vladimir Ivanovich Stepanov's notation system in that Stepanov's system was used for a different genre of dance during a different time period. Stepanov's system was used for ballet after the eighteenth century. The system that Pierre Beauchamp devised was used for Baroque dance until the end of the eighteenth century.

44. The correct answer is C. The passage indicates that Hanva Holm was the first person to register intellectual property rights for a dance notation system. Paragraph 2 states that "Hanva Holm was the first choreographer to copyright the notations of her dance scores." Copyright is a kind of intellectual property right.

45. The correct answer is E. The author most likely mentions Apple and other computerized dance notation systems in the last sentence of the passage in order to indicate possible trends in dance notation. We know this because the sentence focuses on new developments in dance notation.

46. The correct answer is C. The passage suggests that Expressionism illustrates the way in which human beings felt out of sync with their communities at the time this movement was taking place. The last sentence of the passage comments that Expressionism represents "the dislocation of the individual within society."

47. The correct answer is A. The phrase "preponderance of" could be substituted for "lion's share of" in the second paragraph with the least change in meaning. Both phrases refer to the majority of something.

48. The correct answer is B. The primary purpose of the passage is to sum up the historical background and notable features of a particular state. The other answer choices provide specific details from the passage, rather than the primary purpose.

49. The correct answer is D. The discovery that a systematic correspondence exists between the *b* and *v* sounds in certain languages would most strengthen the argument presented in the passage. This correspondence would be similar to the "systematic correspondence between the *p* and *f* sounds in these languages," which is mentioned in the passage.

50. The correct answer is B. The author most likely mentions quantum field theory in order to reveal that the Higgs mechanism inheres in a basic contradiction. In other words, quantum field theory tells us that all particles should be massless, but the Higgs mechanism shows that particles can acquire mass.

51. The correct answer is E. The concept of synchronous rotation, as it is defined in the passage, provides the most likely explanation for the situation in which a telecommunications satellite is always in the same position above a certain city on Earth. This is similar to the way in which the same hemisphere of the Moon always faces the earth.

52. The correct answer is E. Point 5 in paragraph 2 states that the last step in lunar evolution was "the cessation of volcanic activity followed by gradual internal cooling." So, we can conclude that after lunar evolution, the temperature of the internal core of the Moon was lower than it was previously.

53. The correct answer is A. An analysis revealing that there are no geological similarities between samples of material from the surface of the Moon and material from the Earth's crust would tend to disprove the hypothesis that the Moon grew out of debris that was dislodged from the Earth's crust. If we assume that the Moon grew out of material from the Earth, we would expect to see some geological similarities.

54. The correct answer is D. The statement that the wifi area is more popular than the coffee bar is the one that is best supported by the information provided in the chart. According to the chart, the wifi area has a 10% usage rate, but the coffee bar only has an 8% usage rate.

55. The correct answers are B and E. The following groups, when combined, have a utilization rate of more than 15%: Student Union Bar and Meeting Rooms (24% + 5% = 29%) and the Sports Center Sauna and Pool (5% + 11% = 16%)

56. The correct answer is C. The student union has the greatest total utilization rate at Northern College.
The totals per facility are as follows:
Library = 25%
Cafeteria = 18%
Student Union = 29%
Sports Center = 24%
Other Facilities = 4%

Praxis Core Writing Test Format

The Praxis Core Writing Test (5723) consists of three sections: a set of 40 multiple-choice questions and two essay tasks.

Each of the three sections will be timed. You will have 40 minutes to answer the multiple choice questions and 30 minutes to write each of the essays, for a total of 100 minutes to take the entire writing exam.

The 40 multiple choice questions are from the following categories:

- Text types and rhetorical purposes – 6 to 12 questions
- Language and research skills – 28 to 34 questions

The text type and rhetorical purposes questions cover:

- revising and combining sentences
- analyzing word choice, tone, and style

For text type and rhetorical purposes questions, you will see a draft essay, followed by 6 to 8 questions.

You will have to identify how to edit and strengthen the essay, based on the answer choices provided.

Language and research skills questions on the Praxis Core 5723 exam cover:

- using standard English correctly and effectively
- identifying errors and correcting sentences
- recognizing different types of citations
- understanding the use of sources in research
- developing research strategies
- evaluating information for a particular research task

You are not expected to know grammatical terminology on the exam.

However, grammatical terminology is used in this study guide in order to help you review the concepts.

If you are taking the computer-based version of the exam, you will see the multiple-choice questions on your screen.

For the essay part of the test, you will see the essay instructions on the top of the computer screen, and you will compose your response in the lower half of the screen.

Praxis Core Essay Tips

- The essay part of the Praxis Core Writing Exam consists of two essay tasks.
- The first task is an argumentative essay, while the second task is a source-based essay.
- For the argumentative essay, you should be sure to give reasons and examples for the viewpoints that you express.
- For the source-based essays, you will see two passages. You should read both of the texts provided, and then summarize the key points from each one in your written response. Remember to provide citations for your source information in your source-based essay. For instructions on how to provide citations in your essay, please refer to the referencing and citation section later in the study guide.
- The topics for the essays will not require any specialist knowledge in any academic subject area.
- You should read the essay questions and source materials carefully before you begin to construct your responses.
- You should also set aside some time to plan what you are going to say before you begin writing each of the essays.
- Do not attempt to write on a topic other than the ones provided. If you do, you will receive a score of zero.
- You should try to display logical connections among the ideas in your essays.
- You should also write in a variety of sentence patterns, in error-free standard English.
- If you need to brush up on your essay skills, you may want to view the essay correction bonus material at the end of this book before practicing your essay writing with the sample essay questions in this book.

Praxis Core Writing – Grammar Guide

The sections in the following part of the study guide are intended as an overview of the aspects of grammar most commonly tested on the Praxis Core exam.

Read each section carefully, paying special attention to the examples.

Then take the practice writing tests that follow.

Adverb Placement

Adverbs are words that express how an action was done. Adverbs often end in the suffix –ly. You can vary adverb placement, depending upon what you want to emphasize in your sentence. Be sure to place the adverb in the correct position in the sentence and to use the comma, if necessary. If the adverb is used as the first word in a sentence, the adverb should be followed by a comma.

CORRECT: Normally, an economic crisis is a valid reason to raise interest rates.

CORRECT: An economic crisis is normally a valid reason to raise interest rates.

INCORRECT: An economic crisis is a valid reason to normally raise interest rates.

Remember not to place an adverb between "to" and the verb, as in the last example above. This practice, known as the split infinitive, is grammatically incorrect.

Commonly-Confused Words

Be careful with the following commonly-confused words:

- adverse (adjective – detrimental) / averse (adjective – reluctant)
- affect (verb – to cause) / effect (noun – the result or outcome)
- allude (imply) / elude (evade)
- allusion (implication) / illusion (appearance)
- bare (verb – to expose) / bear (verb – to take on a burden)
- bale (noun – a cubed package) / bail (verb – to get something out of something else)
- pore (noun – to read or study with care) / pour (verb – to emit or flow)
- principal (adjective – main or predominant)/ principle (noun – a concept)

Now look at the following examples.

CORRECT: Failure to study will <u>affect</u> your grades.

INCORRECT: Failure to study will <u>effect</u> your grades.

CORRECT: A scientific <u>principle</u> is a concise statement about the relationship of one object to another.

INCORRECT: A scientific <u>principal</u> is a concise statement about the relationship of one object to another.

CORRECT: The run-away thief <u>eluded</u> the police officer.

INCORRECT: The run-away thief <u>alluded</u> the police officer.

CORRECT: He thought he saw an oasis in the desert, but it was an optical <u>illusion</u>.

INCORRECT: He thought he saw an oasis in the desert, but it was an optical <u>allusion</u>.

CORRECT: I was depending on her help, but she <u>bailed</u> out at the last minute.

INCORRECT: I was depending on her help, but she <u>baled</u> out at the last minute.

CORRECT: He <u>pored</u> over the book as he studied for the exam.

INCORRECT: He <u>poured</u> over the book as he studied for the exam.

CORRECT: She is <u>averse</u> to receiving help with the project.

INCORRECT: She is <u>adverse</u> to receiving help with the project.

CORRECT: He could not <u>bear</u> to listen to the loud music.

INCORRECT: He could not <u>bare</u> to listen to the loud music.

Misplaced Modifiers

Modifiers are descriptive phrases. The modifier should always be placed directly before or after the noun to which it relates. Now look at the examples.

CORRECT: Like Montana, Wyoming is not very densely populated.

INCORRECT: Like Montana, there isn't a large population in Wyoming.

The phrase "like Montana" is an adjectival phrase that describes or modifies the noun "Wyoming." Therefore, "Wyoming" must come directly after the comma.

Here are two more examples:

CORRECT: While waiting at the bus stop, a senior citizen was mugged.

INCORRECT: While waiting at the bus stop, a mugging took place.

The adverbial phrase "*while waiting at the bus stop*" modifies the noun phrase "a senior citizen," so this noun phrase needs to come after the adverbial phrase.

Parallel Structure

Correct parallel structure is also known as parallelism. In order to follow the grammatical rules of parallelism, you must be sure that all of the items you give in a series are of the same part of speech. So, all of the items must be nouns or verbs, for example. In other words, you should not use both nouns and verbs in a list. Where verbs are used, they should be in the same form or tense.

CORRECT: The vacation gave me a great chance to unwind, have fun, and experience some excitement. (*Unwind*, *have*, and *experience* are all verbs.)

INCORRECT: The vacation gave me a great chance to unwind, and was fun and quite exciting.

CORRECT: I went jet skiing, surfing, and snorkeling on our vacation. (Skiing, *surfing*, and *snorkeling* are all in the –ing form.)

INCORRECT: I went jet skiing, surfing, and also snorkeled on our vacation.

CORRECT: The hotel was elegant, comfortable, and modern. (*Elegant*, *comfortable*, and *modern* are all adjectives.)

INCORRECT: The hotel was elegant, comfortable, and had up-to-date facilities.

CORRECT: I enjoyed our hotel room, relaxed in the spa, and ate some truly delicious food on our vacation. (*Enjoyed*, *relaxed*, and *ate* are all verbs in the past simple tense.)

INCORRECT: I enjoyed our hotel room, relaxed in the spa, and the food was truly delicious on our vacation.

Pronoun-Antecedent Agreement

Pronouns are words like the following: he, she, it, they, and them. An antecedent is a phrase that precedes the pronoun in the sentence. Pronouns must agree with their antecedents, so use singular pronouns with singular antecedents and plural pronouns with plural antecedents. Be careful not to mix singular and plural forms.

CORRECT: Each student needs to bring his or her identification to the placement test.

INCORRECT: Each student needs to bring their identification to the placement test.

The antecedent "each student" is singular, so the singular pronouns "his" or "her" should follow this antecedent.

CORRECT: The group lost its enthusiasm for the project.

INCORRECT: The group lost their enthusiasm for the project.

The preceding sentence is incorrect because the antecedent is "group," which is singular, while "their" is plural.

Pronoun Usage – Correct Use of *Its* and *It's*

"Its" is a possessive pronoun, while "it's" is a contraction of "it is".

CORRECT: It's high time you started to study.

INCORRECT: Its high time you started to study.

The sentence could also be stated as follows: It is high time you started to study.

Since the contracted form of "it is" can be used in the alternative sentence above, "it's" is the correct form.

CORRECT: A snake sheds its skin at least once a year.

INCORRECT: A snake sheds it's skin at least once a year.

"Its" is a possessive pronoun referring to the snake, so the apostrophe should not be used.

Pronoun Usage – Demonstrative Pronouns

Demonstrative pronouns include the following words: this, that, these, those

"This" is used for a singular item that is nearby. "That" is used for singular items that are farther away in time or space.

SINGULAR: This book that I have here is really interesting.

PLURAL: That book on the table over there is really interesting.

"These" is used for plural items that are nearby. "Those" is used for plural items that are farther away in time or space.

SINGULAR: These pictures in my purse were taken on our vacation.

PLURAL: Those pictures on the wall were taken on our vacation.

Avoid using "them" instead of "those":

INCORRECT: Them pictures on the wall were taken on our vacation.

Pronoun Usage – Relative Pronouns

Relative pronouns include the following: which, that, who, whom, whose

"Which" and "that" are used to describe things, and "who" and "whom" are used to describe people. "Whose" is used for people or things.

WHICH: Last night, I watched a romantic-comedy movie which was really funny.

THAT: Last night, I watched a romantic-comedy movie that was really funny.

WHO: Susan always remains calm under pressure, unlike Tom, who is always so nervous.

"Who" is used because we are describing the person. This is known as the nominative case.

WHOM: To whom should the report be given?

"Whom" is used because the person is receiving an action, which in this case is receiving the report. This is known as the accusative case.

WHOSE: I went out for lunch with Marta, whose parents are from Costa Rica.

WHOSE: I went out for lunch yesterday at that new restaurant, whose name I don't remember.

Please be sure to look at the section entitled "Restrictive and Non-restrictive Modifiers" for information on how to use punctuation with relative pronouns.

Proper Nouns and Proper Adjectives – Capitalization

Proper nouns state the names of specific people, places, ideas, or things. The names of people, countries, states, buildings, streets, rivers, oceans, countries, companies, and institutions are proper nouns. Be careful not to confuse common nouns and proper nouns. Proper adjectives are derived from proper nouns, so they refer to unique classes of people, places, or things. Proper nouns and adjectives should be capitalized. Look at the capitalization in the following examples.

CORRECT: A famous American landmark, the geyser named Old Faithful is located in Yellowstone Park in the northwest corner of the state of Wyoming.

(*American* is a proper adjective. *Old Faithful*, *Yellowstone Park*, and *Wyoming* are proper nouns.)

INCORRECT: A famous american landmark, the geyser named old faithful is located in yellowstone park in the Northwest corner of the State of wyoming.

Punctuation – Using the Apostrophe for Possessive Forms

Apostrophe placement depends upon whether a word is singular or plural.

For the singular, the apostrophe should be placed before the letter "s."

SINGULAR: Our team's performance was poor at the game last night.

For the plural form, the apostrophe should be placed after the letter "s."

PLURAL: Both teams' performances were poor at the game last night.

Remember that the apostrophe is used in sentences like those above in order to show possession. Also remember not to use the apostrophe unnecessarily.

INCORRECT: The date's for the events are June 22 and July 5.

INCORRECT: The dates' for the events are June 22 and July 5.

Punctuation – Using Colons and Semicolons

Colons (:) should be used when giving a list of items. Semicolons (;) should be used to join independent clauses.

COLON: The shop is offering discounts on the following items: DVDs, books, and magazines.

SEMICOLON: I thought I would live in this city forever; then I lost my job.

Note that the word following the semicolon should not be capitalized.

Please see the section entitled "Punctuation and Independent Clauses" for more information on joining clauses.

Punctuation – Using Commas with Dates and Locations

Commas should be used after the date and year in dates. Commas should also be used after towns and states.

DATES: On July 4, 1776, the Declaration of Independence was signed.

LOCATIONS: Located in Seattle, Washington, the Space Needle is a major landmark.

Punctuation – Using Commas for Items in a Series

When using "and" and "or" for more than two items in a series, be sure to use the comma before the words "and" and "or."

> CORRECT: You need to bring a tent, sleeping bag, and flashlight.
>
> INCORRECT: You need to bring a tent, sleeping bag and flashlight.

Notice the use of the comma after the word "bag" and before the word "and" in the series.

> CORRECT: Students can call, write a letter, or send an email.
>
> INCORRECT: Students can call, write a letter or send an email.

Notice the use of the comma after the word "letter" and before the word "or" in the series.

Punctuation and Independent Clauses – Avoiding Run-On Sentences

Run-on sentences are those that use commas to join independent clauses together, instead of correctly using the period.

Because they incorrectly use the comma to fuse sentences together, run-on sentences are sometimes called comma splices.

An independent clause contains a grammatical subject and verb. It therefore can stand alone as its own sentence.

The first word of the independent clause should begin with a capital letter, and the clause should be preceded by a period.

> CORRECT: I thought I would live in this city forever. Then I lost my job.
>
> INCORRECT: I thought I would live in this city forever, then I lost my job.

"Then I lost my job" is a complete sentence. It has a grammatical subject (I) and a verb (lost).

The independent clause must be preceded by a period, and the first word of the new sentence must begin with a capital letter.

Alternatively, an appropriate conjunction can be used to join the independent clauses:

> CORRECT: I thought I would live in this city forever, and then I lost my job.

Restrictive and Non-restrictive Modifiers

Restrictive modifiers are clauses or phrases that provide essential information in order to identify the grammatical subject. Restrictive modifiers should not be preceded by a comma.

> CORRECT: My sister who lives in Indianapolis is a good swimmer. (The speaker has more than one sister.)

In this case, the speaker has more than one sister, and she is identifying which sister she is talking about by giving the essential information "who lives in Indianapolis."

On the other hand, a non-restrictive modifier is a clause or phrase that provides extra information about a grammatical subject in a sentence. A non-restrictive modifier must be preceded by a comma. Non-restrictive modifiers are also known as non-essential modifiers.

> CORRECT: My sister, who lives in Indianapolis, is a good swimmer. (The speaker has only one sister.)

In this case, the speaker has only one sister. Therefore, the information about her sister's city of residence is not essential in order to identify which sister she is talking about. The words "who lives in Indianapolis" form a non-restrictive modifier.

Sentence Fragments

A sentence fragment is a group of words that does not express a complete train of thought.

CORRECT: I like Denver because it has a great university.

INCORRECT: I like Denver. Because it has a great university.

In the second example, "because it has a great university" is not a complete thought. This idea needs to be joined with the previous clause in order to be grammatically correct.

Subject-Verb Agreement

For questions on subject-verb agreement, you need to be sure that subjects agree with verbs in number. In other words, use a singular verb with a singular subject and a plural verb with a plural subject. While this sounds straightforward, complications can arise with certain words like "each," "every," "neither," and "either," all of which are in fact singular. Subject-verb agreement can also be confusing when there are intervening words in a sentence.

CORRECT: The flowers in the pots in the garden grow quickly.

INCORRECT: The flowers in the pots in the garden grows quickly.

The grammatical subject in the above sentence is "flowers," not "garden," so the plural form of the verb (*grow*) needs to be used.

CORRECT: Each person in the groups of students needs to pay attention to the instructions.

INCORRECT: Each person in the groups of students need to pay attention to the instructions.

The grammatical subject in the above sentence is "each person," not "students." "Each" is singular and therefore requires the singular form of the verb (*needs*).

CORRECT: Each of the men is very strong and determined.

INCORRECT: Every one of the books are on the shelf.

Subordination

Subordinators include words and phrases such as "although," "but," "even though," "because of," and "due to." Be careful to use commas correctly when subordinating sentences.

CORRECT: I was going to study this evening, but the noise next door made it impossible.

INCORRECT: I was going to study this evening but the noise next door made it impossible.

CORRECT: Although I was going to study this evening, the noise next door made it impossible.

INCORRECT: Although I was going to study this evening the noise next door made it impossible.

The word "but" is a subordinator. Subordinators need to be preceded by a comma, so the first sentence is correct as written.

You also need to use a comma in the middle of the sentence when beginning the sentence with the subordinator.

Review of Verb Tenses

Present simple tense

The present simple tense is used for habitual actions.

Example: He *goes* to the office at 8:00 every morning.

The present tense is also used to state scientific truths, facts, and generalizations.

Example: Water *freezes* at zero degrees Celsius.

Past simple tense

The past simple tense is used for actions that were started and completed in the past.

Example: I walked three miles yesterday.

Present perfect tense

The present perfect tense is used for actions that were completed in the past, but that have relevancy in the present time.

Example: I *have studied* every day this week.

The phrase "this week" shows that the action has relevancy in the present time.

Passive tenses

Use the passive voice to emphasize the object of the action, rather than the person who was conducting the action.

In the example sentence that follows in this section, the diplomas are the object of the action.

We want to emphasize the fact that students are receiving the diplomas. We want to de-emphasize the fact that the university officials are the people responsible for handing out the diplomas.

Example: Diplomas *are handed out* on graduation day every year.

The passive can also take more complicated forms.

Example: Ronald Reagan was the only former actor *to have been elected* President in the twentieth century.

The general populace of the United States elects the president. We want to emphasize who was elected, rather than who did the electing, so we need to use the perfect form of the passive in the preceding sentence.

Praxis Core Writing – Source, Reference, and Citation Guide

Sources

Be sure that you know the difference between primary and secondary sources for your Praxis Core Writing Test.

Primary Sources – original information in the form of a text, document, manuscript, survey, or statistical data.

Examples: Archives at the county court house on births and deaths

Records held by an institution

A literary manuscript produced by the author

Secondary sources – commentary, discussion, or analysis of the primary source.

Examples: An independent research report analyzing births and deaths in the county

A newspaper article that comments on the records held by the institution

An article in a scholarly journal that discusses the literary manuscript

Use of sources

Be sure that you know how to use lists of references in your own research. You will most likely encounter an exam question on this research technique.

When you are conducting research for a dissertation, term paper, or report, you should pay special attention to the list of references cited at the end of the book or article.

You can read the sources provided in the list of references, and then identify and note down the names of interesting-looking books and articles that the author has cited. You can then read these sources later and cite them in your own research.

Referencing

You will need to be able to identify what type of source is being referenced when you look at a citation. Note that citations vary in format, depending on what type of material is being used. Study the following citations and notice the slight differences in format for each one.

Books:

Brown, Mark. (2012). *Old towns: A study in urban development.* Pittsburg: Authors' Press.

Notice that only the year is given for a book. Also note that the title is given in italic font. The place and company of publication are provided at the end of a book citation.

Academic or Scholarly Journal:

Tauton, Rachel. (2009). The use of audio in the English classroom. *Pedagogy and Learning, 23*(4), 13-21.

Notice that the volume and issue numbers [*23*(4)] are provided for a scholarly journal. In addition, the page numbers are given at the end of the citation in the following format: 13-21. The name of the journal is given in italics, but the title of the article is not italicized.

Magazines:

Haas, Assan. (2012, May 9). Power in the new millennium. *Popular Technology,* 135, 28-31.

Notice that the date for a magazine citation is in the following format: (2012, May 9). The issue number will be given (135), followed by the page numbers (28-31). The name of the magazine is given in italics, but the title of the article is not italicized.

Newspapers:

Gomez, Joaquin. (2015, January 28). The cost to the taxpayer of state energy policies. *The News Today*, pp. 2B, 3A.

The date for a newspaper citation is in the same format as that of a magazine: (2012, May 9). However, the page numbers for a newspaper state the section and page number in this format: pp. 2B, 3A. The name of the newspaper is given in italics, but the title of the article is not italicized.

Online sources:

Papadopoulos, Maria. (2012). Twenty top study tips. *Lists for Busy People.* Retrieved from http://www.alistapart.com/articles/lists

A URL or web address needs to be provided for citations of online materials. The name of the website is given in italics, but the title of the article is not italicized.

Citing works in your source-based essay:

As you write your source-based essay, remember to use the author's name when discussing his or her ideas. You will also need to use the author's name and title of the article when you give an exact quotation form the source.

Look at the following example:

Source: Lewis, Lamar. (2014). DNA Research and Gene Splicing. *Medicine and Technology, 15*(2), 35-42.

> While some horticulturalists have been adept at and accustomed to using gene splicing for decades in the form of creating hybrids of plants, the subject of using gene splicing in DNA research has been surrounded by controversy lately. As animal rights groups have come into prominence socially and politically, people are increasingly aware of the suffering of animals. Thus, a growing contingent of the population is questioning whether using animals for DNA research is medically reasonable and ethical.

Excerpt from source-based essay:

Lewis states that gene splicing has been in use for a few years already, in the form of the creation of plant hybrids. Yet, he points out that some groups have concerns about this practice, particularly about the manner in which animals are treated during the research process. He asserts that "a growing contingent of the population is questioning whether using animals for DNA research is medically reasonable and ethical" (DNA Research and Gene Splicing).

The essay has correctly cited the source because:

- The author's name (Lewis) is given at the beginning of the paragraph.
- The student synthesizes the most important information from the source and uses his or her own words in sentences 1 and 2 from the essay.
- When the exact wording has been taken from the source in sentence 3, the words are enclosed in quotation marks and the name of the article is provided in parentheses at the end of the sentence.

Praxis Core Writing Practice Test 1

Instructions for Questions 1 to 19 – Each of the sentences below has four underlined parts. Read each sentence and determine whether any of the underlined parts contains an error in grammar or use. If so, select the underlined part that contains the error as your answer. If the sentence contains no error, then select "No error." No sentence contains more than one error.

1. One of the first skyscrapers to be erected in New York, at the site where its television mast would be most effective, the Empire State Building was constructed in the Art Deco style around 1930. No error

2. Even though chicory is usually cooked, their leaves can be eaten in various ways, including as a raw ingredient in salads or as a dried and ground substitute for coffee. No error

3. Many composers promote nationalism in their work, but what is different about Mussorgsky's compositions are the overt patriotism in his operas. No error

4. Lymph glands, masses of tissue that form part of the lymphatic system, filter bacteria and organisms from organ's in the body and allow lymph to flow into capillaries and from them into lymph vessels. No error

5. Scientifically speaking , almost any positive electric charge can produce a unit of energy from which a volt of electricity is derived. No error

6. It is difficult to know when people first began to grow sweet potatoes, because early settlers often did not differentiate this vegetable to other tubers. No error

7. Although companies try to price their products competitively, they often cannot , in spite of their best plans, introduce no new merchandise into the market at a profit. No error

8. Commencing in 1847 and for nearly 16 years thereafter, Mexico City was occupied by U.S. troops until it was conquered by Maximilian, a ruler whose name originates from the roman word *maximus*. No error

9. Ratified at the end of 1992 , the North American Free Trade Agreement eliminated trade barriers among Canada and the United States in order to increase trade and to enable each of the counties to find its unique market position. No error

10. Scientists have discovered that most comets in orbit around the sun , seem to be composed of rock and dust particles embedded in ice. No error

11. Since Christopher Columbus believed that he had established a route between the East Indies and China, subsequent explorer's journeys were impeded by all other theories requiring them to think otherwise. No error

12. Some residents of East Germany opposed the dismantling of the Berlin Wall, while others despaired of its very existence because many residents die when attempting to cross it. No error

13. Originally focusing on suffrage, the Women's Rights Movement expanded to include much more than the right to vote and effected the role of women in many countries around the world. No error

14. The Inca migration from the Peruvian highlands to the area west of the Andes constitute an example of the consolidation and extension of the tribe's empire in South America. No error

15. When a government establishes a public sector borrowing requirement, it raises money through the issuance of stocks and bonds, which not only increases its available funds and also forms part of the national debt. No error

16. Ceded by the British to the United States in 1783, the Territory of Wisconsin was inhabited by settlers who differed from that of other territories, not settling the region in haste as newcomers to other states did. No error

17. Born in Hamlet , North Carolina, John Coltrane was arguably one of the most famous of all jazz musicians having played the saxophone. No error

18. A rope and pulley system, an everyday apparatus that easily lifts heavy items, having long been recognized by physicists as a useful application of applied force. No error

19. Great Danes, large hunting dogs that were bred originally by Germans, were once the principle companion animal for the wealthy. No error

Instructions for Questions 20 to 30 – Select the best substitute for the **highlighted** parts of the following ten sentences. The first answer [choice A] is identical to the original sentence. If you think the original sentence is best, then choose A as your answer.

20. The name "catfish" is applied to a very large family of freshwater fish called the Siluriformes, which have whisker-like barbels **growing from its mouth**.
   A. growing from its mouth.
   B. growing from it's mouth.
   C. growing from their mouths.
   D. grown from the mouth.
   E. to grow from the mouth.

21. While driving the car this afternoon, **the tire suddenly went flat and I was surprised.**
   A. the tire suddenly went flat and I was surprised.
   B. the tire suddenly went flat, and I was surprised.
   C. I was surprised when the tire suddenly went flat.
   D. it was a surprise when the tire suddenly went flat.
   E. it was a surprise as the tire suddenly went flat.

22. Best known for his science fiction classic *Star Wars*, US film producer and director George Lucas was revolutionary at the time for having worked with Francis Ford Coppola, retaining the merchandising rights for his works, **and the special effects to be used in his movies.**
   A. and the special effects to be used in his movies.
   B. and using special effects in his movies.
   C. and the special effects used in his movies.
   D. and the special effects he used in his movies.
   E. and used special effects in his movies.

23. We were planning a picnic for the kids in the park **this afternoon, the rain made it impossible.**
   A. this afternoon, the rain made it impossible.
   B. this afternoon, the rain making it impossible.
   C. this afternoon, with the rain made it impossible.
   D. this afternoon, but the rain made it impossible.
   E. this afternoon, it was the rain that made it impossible.

24. It is said that Roosevelt's favorite pursuits were **to read, play games, and doing crossword puzzles.**
   A. to read, play games, and doing crossword puzzles.
   B. to read, games, and crossword puzzles.
   C. reading, gaming and do crossword puzzles.
   D. Reading, games, and doing crossword puzzles
   E. reading, playing games, and doing crossword puzzles.

25. **Being frustrated and fed up, and tired from a hard week at work,** she decided it was time to go away on vacation.
   A. Being frustrated and fed up, and tired from a hard week at work,
   B. Being frustrated and fed up, and she was tired from a hard week at work,
   C. With being frustrated and fed up, and tired from a hard week at work,
   D. Frustrated, fed up, and tired from a hard week at work,
   E. Frustrated and fed up, and also tired from a hard week at work,

26. With his unconventional style, E.E. Cummings inspired a generation of experimental artists called "avant-garde poets," **who disregarded poetic form and rebelling** against the use of traditional spelling and punctuation in verse.
A. who disregarded poetic form and rebelling
B. who disregarding poetic form and rebelling
C. who disregarded poetic form and rebelled
D. to disregard poetic form and rebelling
E. who, disregarding poetic form to rebel

27. An outstanding military commander and statesman, George Washington **having strictly avoided** overstepping the constitutional limitations of presidential power.
A. having strictly avoided
B. having to strictly avoid
C. has strictly avoided
D. was strictly avoiding
E. strictly avoided

28. Dolomite, otherwise known as calcium-magnesium carbonate, is one of many minerals in **sedimentary rock forming crystals** and to be used for ornamental purposes.
A. sedimentary rock forming crystals
B. sedimentary rock formations of crystals
C. sedimentary rock to form crystals
D. sedimentary rock having formed crystals
E. sedimentary rock that had formed crystals

29. **Despite the fact he worked overtime several days in a row the congressman** didn't finish his report on time.
A. Despite the fact he worked overtime several days in a row the congressman
B. Despite the fact that he worked overtime several days in a row, the congressman
C. In spite of the fact working overtime several days in a row, the congressman
D. Although overtime was worked several days in a row, the congressman
E. He worked overtime several days in a row, however the congressman

30. **If text messaging while at the wheel,** you could be charged with dangerous driving in some states.
A. If text messaging while at the wheel,
B. While text messaging at the wheel,
C. If you text message while at the wheel,
D. If you text messaging while at the wheel,
E. Text messaging at the wheel

Instructions for Questions 31 to 36 – Read the draft essay below and then choose the best answers to the questions that follow.

(1) A group of English separatists known as the Pilgrims left England. ***(2)*** When they left England, they went to live in Amsterdam in 1608. ***(3)*** After spending a few years in their new city however many members of the group felt that they did not have enough independence. ***(4)*** In 1617, the Pilgrims decided to leave Amsterdam to immigrate to America.

(5) Due to their lack of social standing, they had many financial problems that prevented them from beginning the journey. ***(6)*** Their inability to finance themselves caused many disputes and disagreements. ***(7)*** The Pilgrims finally managed to resolve these conflicts when they obtained financing from a well-known and respected London businessman named Thomas Weston.

(8) Having secured Weston's monetary support, the group returned to England to pick up some additional passengers. ***(9)*** After 65 days at sea, the Pilgrims reached America in December 1620. ***(10)*** The early days of their new lives were filled with hope and promise. ***(11)*** These hopes were dashed when the harsh winter proved to be too much for some of the settlers. ***(12)*** Nearly half of the Pilgrims died during that first winter.

31. What is the best way to revise and combine sentences 1 and 2?

A. A group of English separatists known as the Pilgrims went to live in Amsterdam in 1608 after they left England.
B. A group of English separatists known as the Pilgrims left England and went to live in Amsterdam in 1608.
C. A group of English separatists known as the Pilgrims left England, they went to live in Amsterdam in 1608.
D. A group of English separatists known as the Pilgrims left England since they went to live in Amsterdam in 1608.
E. A group of English separatists known as the Pilgrims left England, since they went to live in Amsterdam in 1608.

32. What is the best to do to sentence 3? Sentence 3 is provided again here for ease of reference.
After spending a few years in their new city however many members of the group felt that they did not have enough independence.
A. Put commas after the word "city" and after the word "however."
B. Replace the words "did not" with the contraction "didn't."
C. Place a comma after the word "years."
D. Replace the word "spending" with the words "they spent."
E. Delete it from the passage.

33. Which one of the following words or phrases is best to insert at the beginning of sentence 4?
A However,
B. In addition,
C. Therefore,
D. Surprisingly,
E. In fact,

34. Which of the following is the best version of the underlined portion of sentence 5? Sentence 5 is provided again here for ease of reference.
Due to their lack of social standing, <u>they had many financial problems that prevented them from beginning the journey</u>.
A. it was said that they had many financial problems that prevented them from beginning the journey.

B. the members of the group experienced many financial problems that prevented them from beginning the journey.
C. they had many economic problems that prevented them from beginning the journey.
D. they had many financial problems that prevented them from beginning.
E. they had many financial problems that prevented them.

35. Which of the following sentences would be best inserted between sentences 8 and 9?
A. New evidence indicates that the Pilgrims suffered from many diseases that would have been treatable had they had access to modern-day antibiotics.
B. Some scholars argue that life on the *Mayflower* was filled with contention and in-fighting.
C. The life of Pocahontas followed a completely different trajectory.
D. They boarded a large ship called the *Mayflower* on September 16, 1620.
E. Long sea voyages were not unheard of in those days.

36. Which one of the following words or phrases would be best inserted at the beginning of sentence 12? Sentence 12 is provided again here for ease of reference.
Nearly half of the Pilgrims died during that first winter.
A. Ironically,
B. Unfortunately,
C. Finally,
D. To conclude,
E. Summing up,

Instructions for Questions 37 to 40 – Read the basic research skills questions that follow and then choose the best answers.

37. Which of the following best describes the term "secondary source"?
    A. physical evidence from the time under study
    B. a document that was written during the time under study
    C. a source that analyzes or discusses information originally published elsewhere
    D. a compendium of knowledge, such as an encyclopedia
    E. a source that provides a broad introductory overview on the topic

38. Anderson, G. (2015, January 7). Neighborhood decline in American cities. *Society Today*, 125(3), 25-32.
    The item above is from a list of references. Which of the following is being cited?
    A. an article from a scholarly journal
    B. a blog
    C. a newspaper article
    D. a magazine article
    E. a book

39. A student is conducting research on possible interpretations of a new federal regulation on carbon emissions from automobiles. Which of the following pieces of information would NOT be directly relevant to the student's research?
    A. Excerpts of text from the new federal regulation
    B. Graphs and other visual data from reports on the new federal regulation
    C. Statistical evidence of the way in which previous federal regulations on carbon emissions have been prone to misinterpretation
    D. Interviews with politicians who discuss how they would interpret the new federal regulation
    E. A scholarly article which concludes that many state and local laws have had various possible interpretations

40. Which of the following is a primary source on the effects of climate change in North America?
    A. An article from a scholarly journal that summarizes previous studies on the subject of global climate change
    B. A scientific study that provides evidence of the effects of climate change on North American cities.
    C. A survey of local residents' opinions on new laws that regulate climate change
    D. A detailed analysis of the pollutants emitted from North American factories
    E. A blog that discusses climate change in North America.

Praxis Core Writing Practice Test 1 – Answers and Explanations

1. You should select: No error. The sentence contains no errors in grammar or usage.

2. You should select: their. The antecedent is "chicory," which is singular. So, the pronoun "it" should be used before the grammatical subject, rather than "their."

3. You should select: compositions are. The grammatical subject of this clause is the word "what." Therefore, we need to use the singular form of the verb (is).

4. You should select: organ's. The plural form "organs" is needed. The possessive form "organ's" is incorrect because there is no corresponding noun for the possessive.

5. You should select: No error. The sentence contains no errors in grammar or usage.

6. You should select: to. The verb "differentiate" takes the preposition "from", not "to."

7. You should select: no. The sentence contains a double negative since the word "cannot" is used earlier in the sentence. So, "any" should be used here.

8. You should select: roman. Roman is a proper noun, so it needs to be capitalized.

9. You should select: among. Only two counties are mentioned, so the word "between" is needed.

10. You should select: the comma. The comma is incorrect since it unnaturally divides the subject and verb of the clause.

11. You should select: explorer's. The second part of the sentence is describing more than one explorer since the word "them" is used later in the sentence. Therefore, the plural possessive is needed (explorers').

12. You should select: die. The sentence is written in the past simple tense, so the verb form "died" should be used.

13. You should select: effected. The verb form (affected) is required since the sentence is talking about the changes that were a result of the Women's Rights Movement.

14. You should select: constitute. The subject of the sentence is "migration" so the singular verb form (constitutes) should be used.

15. You should select: and. We can see the use of "not only" earlier in the sentence, so we need "but also," instead of "and also"

16. You should select: that. The subject of this part of the sentence is the plural "settlers," so the plural pronoun "those" is needed, rather than "that."

17. You should select: having played. The perfect infinitive verb form "to have played" is needed because we are talking about a topic that has current relevance. We know that the topic is under current discussion because of the word "arguably" earlier in the sentence.

18. You should select: having. The subject of the sentence is "system," so the present simple verb form "has" is needed.

19. You should select: principle. You should use the word "principal" as an adjective in this sentence, rather than the noun form "principle."

20. The correct answer is C. The antecedent is "Siluriformes," which is plural. However, "its mouth" is singular. Therefore, the plural pronoun "their" and the plural noun "mouths" need to be used in order to correct the error.

21. The correct answer is C. Because of the placement of modifying phrase at the beginning, the sentence as written suggests that the tire is driving the car. The pronoun "I" needs to be placed after the introductory phrase since the speaker was driving the car. Remember to place the modifying phrase directly before the noun that it modifies.

22. The correct answer is B. The sentence contains a lack of parallel structure ("for having worked [. . .], retaining [. . .], and the special"). Answer B correctly uses the –ing form in all three verbs in order to correct the error.

23. The correct answer is D. The sentence as written is an example of a run-on sentence. The word "but" correctly subordinates the second part of the sentence.

24. The correct answer is E. The sentence has an error in parallelism. All of the hobbies should be stated in the –ing form, so E is the correct response.

25. The correct answer is D. This sentence also has an error in parallelism. All of actions should be stated in the –ed (past simple) form, so D is the correct response.

26. The correct answer is C. The sentence has an error in parallel construction, since it uses the –ed form of the verb ("disregarded") and the –ing form of the verb ("rebelling") in the same part of the sentence. There is no problem with the relative pronoun usage ("who") in the original sentence. Answer C correctly uses "who" and constructs both verbs in the –ed form

27. The correct answer is E. The sentence as written is a fragment. Replacing the –ing form "having" with the past simple tense ("avoided") completes the sentence correctly since the action described in the sentence occurred in the past.

28. The correct answer is C. The sentence as written incorrectly implies that the rocks, rather than the minerals, form the crystals. Response C corrects this error by using the verb "to form."

29. The correct answer is B. The highlighted part of the sentence is missing a relative pronoun ("that"). In addition, a comma needs to be placed before the grammatical subject of the sentence ("the congressman").

30. The correct answer is C. Since the pronoun "you" is used in the second part of the sentence, the same pronoun is also required in the first part of the sentence. The sentence is describing a generalization, so the present simple tense ("message") is needed, rather than the continuous form ("messaging").

31. The correct answer is B. The sentences need to be grammatically correct and the clauses should be in their original order in order to give the correct emphasis to the ideas.

32. The correct answer is A. The word "however" needs to be offset with commas when used in the middle of a sentence like this one. B and D are possible answers, but not the best answers. B is not the best answer because full forms like "did not" are more common in academic writing than contractions. D is not the best answer because of the repetition of the pronoun "they." Answer C suggests an incorrect comma placement.

33. The correct answer is C. Sentence 3 talks about the lack of independence the Pilgrims felt, which was the reason they immigrated to America. "Therefore" is the only answer choice that implies a cause-and-effect relationship between the two sentences, so it is the correct answer.

34. The correct answer is B. Replacing the pronoun is the best choice for stylistic reasons in this instance. The sentence already contains the pronouns "their" and "them." Accordingly, the replacement of the pronoun "they" makes the sentence less repetitious.

35. The correct answer is D. Sentence D fits in with the chronological flow of events mentioned in the paragraph. The other answers do not.

36. The correct answer is B. This is a question on logical sentence linking. We are mentioning a negative, unforeseen outcome, so "unfortunately" is the best answer.

37. The correct answer is C. Secondary sources provide commentary on primary sources. Refer back to the referencing and citation section of the study guide if you are still unsure about the difference between primary and secondary sources.

38. The correct answer is A. The indication of a volume and issue number ["125(3)"] indicates that a scholarly journal is being cited.

39. The correct answer is E. Answers A, B, C, and D address the new federal regulation or previous federal regulations on the subject of carbon emissions. Answer E applies to states and local laws, rather than federal regulations, so this source would not be relevant to the student's research.

40. The correct answer is B. Answers A, C, and E are secondary sources on the subject of climate change. Answer D is a primary source on pollution, rather than climate change itself, so it is slightly off the point. A scientific study is one type of primary source, so answer B is the correct response.

Argumentative Essay Topic – Practice Test 1

You may wish to refer back to the section entitled "Praxis Core Essay Tips" before writing the essays in this part of the practice test.

Instructions: You are allowed 30 minutes to complete each essay in this part of the test. When you have finished, study the model essays and comments that follow each sample response.

Look at the essay topic below and then write your essay.

Thomas Edison once stated: "Restlessness and discontent are the first necessities of progress."

To what extent do you agree or disagree with the above statement? Provide reasons and examples to support your response.

Sample Argumentative Essay – Practice Test 1

Like Thomas Edison, I support the view that restlessness and discontent are the first necessities of progress. This simple assertion holds true both on the societal and personal levels.

It is irrefutable that restlessness can lead to progress within society. During the pre-revolutionary period in American history, for example, the settlers in the American colonies became very restless with the way that English law was treating them. This restlessness led to the American Revolutionary War, which witnessed the birth of a myriad of personal and social liberties that American citizens still enjoy today.

In addition, great innovations have come about and continue to come about because of discontent or dissatisfaction with the current state of affairs. Because of this basic principle, many great inventions have been created and many discoveries have occurred. In the nineteenth century, for instance, Louis Braille had an accident at three years of age which caused him to become blind. When he became older, Braille realized that the vast intellectual world of thought and ideas would be closed to him forever unless he devised a system whereby the blind could read. It was this dissatisfaction that led Braille to create the system of type that sight-impaired people around the world utilize today.

Likewise, restlessness and discontent on an individual level can also lead to personal progress. I myself have had life experiences that illustrate this principle. Having worked in an office for many years as an accountant, I realized that I was restless with and discontent in this line of work. This dissatisfaction led me to a journey of self-discovery, culminating in my decision to return to college as a mature student in order to study education.

That is not to say that satisfaction and contentment, whether on a personal or social level, are not to be sought-after. But while satisfaction and contentment can be admirable characteristics in certain ways, these states of mind rarely lead to the social or personal struggles that are necessary in order for change or innovation to occur.

Comments on the Sample Argumentative Essay for Practice Test 1

Like Thomas Edison, I support the view that restlessness and discontent are the first necessities of progress. This simple assertion holds true both on the societal and personal levels.

It is irrefutable that restlessness can lead to progress within society. **During** the pre-revolutionary period in American history, **for example**, the settlers in the American colonies became very restless with the way that English law was treating them. This restlessness led to the American Revolutionary War, **which** witnessed the birth of a myriad of personal and social liberties that American citizens still enjoy today.

In addition, great innovations have come about and continue to come about because of discontent or dissatisfaction with the current state of affairs. **Because of** this basic principle, many great inventions have been created and many discoveries have occurred. In the nineteenth century, **for instance**, Louis Braille had an accident at three years of age which caused him to become blind. When he became older, Braille realized that the vast intellectual world of thought and ideas would be closed to him forever unless he devised a system **whereby** the blind could read. It was this dissatisfaction that led Braille to create the system of type that sight-impaired people around the world utilize today.

Likewise, restlessness and discontent on an individual level can also lead to personal progress. I myself have had life experiences that illustrate this principle. Having worked in an office for many years as an accountant, I realized that I was restless with and discontent in this line of work. This dissatisfaction led me to a journey of self-discovery, culminating in my decision to return to college as a mature student in order to study education.

That is not to say that satisfaction and contentment, whether on a personal or social level, are not to be sought-after. **But while** satisfaction and contentment can be admirable characteristics in certain ways, these states of mind rarely lead to the social or personal struggles that are necessary in order for change or innovation to occur.

Comments: This essay would receive a high score because it fulfils each of the six grading criteria provided below.

1) Clear thesis statement idea – the writer expresses his central idea clearly in the opening paragraph of the essay when he states: "I support the view that restlessness and discontent are the first necessities of progress [. . . which] holds true both on the societal and personal levels."

2) Well-supported – the essay is well-supported with meaningful explanations and examples. The writer gives examples from Louise Braille's life and the American Revolutionary War, as well as a description of a personal experience to illustrate his viewpoint.

3) Effective organization – the essay is structured very well and is easy to follow and read. The arguments in support of the social implications of the assertion are stated in paragraphs 2 and 3, while the personal implications of the assertion are stated in paragraph 4.

4) Grammatical conventions – the student has utilized correct grammar, spelling, and punctuation. The essay is also organized into meaningful paragraphs, each of which centers around a key example or point.

5) Sentence patterns – the structure of the sentences in this essay is very complex. The writer uses linking words and phrases such as *when*, *in addition*, *during*, and *likewise* to achieve this effect. Please study the words and phrases highlighted in **bold** in the essay and pay attention to how they are being used.

6) Appropriate tone and style – the writer expresses his points of view in a formal, academic way. The level of vocabulary he uses demonstrates that he is writing to an audience of educated adults.

Source-Based Essay Task – Practice Test 1

Directions:

When paraphrasing or providing quotations from the source below, provide a citation for each source by referring to the author's last name, the title of the source, or any other clear identifying method.

Assignment:

Both of the following passages address the topics of organic farming and consumers' views of organic food. Read the two sources carefully and then write an essay in which you identify the most important points from the passages.

<u>Source 1:</u>

Walton, Tamsin. *Organic Food and the Views of the American Consumer*, NY: Newton University Press. 2015. Print.

Even though organic farming and organic produce create many positive outcomes for the environment, some mainstream American consumers have reservations about organic food.

The first drawback that consumers perceive is perhaps the most obvious: cost. Organic food often costs fifty to one hundred percent more than food produced using traditional farming methods. Consumers with higher income levels can probably afford this, but many people simply do not believe that the potential health and environment benefits are worth the expense.

There are also concerns about the safety of organic food. Organic produce is often grown using cow manure. Take the case of windfall apples, which are apples that fall off the tree onto the ground below. These apples can be contaminated by cow manure, and if not washed properly before human consumption, serious food poisoning or even death can occur. This is because manure contains a very dangerous form of bacteria known as e-coli.

The organic food industry is quick to point out, of course, that instances of food poisoning are extremely rare. This assertion is true, but since the news media heavily publicizes cases of e-coli infection, people are still very reluctant about buying organic food for this reason.

Last but not least, and strangely enough, some people do not like organic food because they think it spoils too quickly. Food preservatives are not natural ingredients in food, but they do substantially prolong the life of food in many cases. The longer life of the product makes the non-organic food a better value in the minds of many consumers. Therefore, it may be quite some time before the purchase of organic food becomes the norm in American households.

<u>Source 2:</u>

Spencer, Franklin. *Organic Farming: The Monetary and Ecological Benefits*. NJ: Brownville Press. 2014. Print.

Organic farming has become one of the fastest growing trends in agriculture recently. Over the past ten years, sales of organic products in the United States have increased a staggering twenty percent, with retail sales per year of more than nine billion dollars. American farmers have realized that organic farming is an incredibly cost-effective method because it can potentially be used to control costs, as well as to appeal to higher-priced markets.

Apart from these monetary benefits, organic farming also naturally results in positive ecological outcomes for the environment. Organic farming relies on practices that do not harm the environment, and for this reason, chemicals and synthetic medicines are prohibited. All kinds of agricultural products can be produced organically, including grains, meat, eggs, and milk.

In order for agricultural products to be certified as organic, they must be grown and processed according to regulations established by the USDA, the United States Department of Agriculture. Certification involves two stages: the submission of a system plan and an inspection of processing facilities. The certification process is a stringent one and must be undertaken every year.

In spite of these rigorous requirements, some people remain concerned about the safety of organic food. However, research has shown that organic produce contains lower levels of both chemicals and bacteria than food which is produced using conventional farming methods.

Last but not least, organic farms are better for wildlife that those run conventionally. Scientists have discovered that organic farms contain more species of plants, birds, and insects due to the fact that the absence of chemicals from pesticides and fertilizers makes these areas richer habitats for animals.

Sample Source-Based Essay – Practice Test 1

Organic food has become a popular trend in food production in the United States. **Indeed**, statistics show that more than nine billion dollars' worth of organic food is purchased each year (Spencer). **However**, like most debatable topics, the subject of organic food has both *protagonists* and *antagonists*.

Spencer points out that organically farmed food provides *monetary* benefits to farmers, **who** can use this production method to control their costs and then charge *consumers* a higher price. **He adds that** organic food production is **also** better for the environment than traditional farming methods, **particularly** for the *diversity* of wildlife. Therefore, "organic farms contain more species of plants, birds, and insects due to [. . .] the absence of chemicals from pesticides" (*Organic Farming: The Monetary and Ecological Benefits*). **Spencer also explains that** organic farming is regulated by the United States Department of Agriculture, **meaning that** organic products must conform to *stringent* regulations.

Nevertheless, organic farming also has its *detractors*. In her book on organic food and the American consumer, **Walton asserts that** some people feel that organic food is simply too expensive. **She states that** some consumers **have expressed concerns that** organic food might be harmful to their health because it may be prone to bacterial *contamination*. But perhaps her most unexpected indictment is that "strangely enough, some people do not like organic food because they think it spoils too quickly" (*Organic Food and the Views of the American Consumer*).

In conclusion, although there is disagreement about whether the advantages of organic food outweigh the disadvantages, **the areas of debate on this topic revolve primarily around** three issues: cost, safety, and balancing the benefits of organic food to the individual against positive outcomes of organic farming for the environment.

Comments on Sample Source-Based Essay 1

The essay above is high level for the following main reasons:

1) It summarizes the most significant points from both of the source passages.

2) The transitions from source information to the student's own paraphrasing are smooth, making the essay flow well.

3) The student has provided citations for each use of the references, as required.

4) The essay is well organized. Notice the words and phrases in **bold**, which the student uses to introduce the main points from the sources and to make the essay flow well.

5) The essay is grammatically correct. Notice that in summaries for the source-based writing task, the verb in the opening sentence of the essay is often given in the present perfect tense. The verbs in the other sentences should be in present simple, active or passive.

6) The student adds her own explanations to the sources, using high-level academic vocabulary. Please notice the advanced vocabulary words in *italics*. Try to learn these words if you do not know them already.

Praxis Core Writing Practice Test 2

Instructions for Questions 1 to 19 – Each of the sentences below has four underlined parts. Read each sentence and determine whether any of the underlined parts contains an error in grammar or use. If so, select the underlined part that contains the error as your answer. If the sentence contains no error, then select "No error." No sentence contains more than one error.

1. Geologists have noted that petroleum is a chemical compound consisting of a complex mixture of hydrocarbons , that appear to be formed from the bodies of long-dead organisms. No error

2. Magnetism occurs when a current is associated with a field of force and with a north-south polarity, which means that any substance tends to align themselves with the field. No error

3. Charlemagne was crowned King of the Franks in 768, and for approximately 46 years afterwards, he engaged in a brutal conquest of Europe, including the campaign against the moorish people in Spain. No error

4. While both the Black Mountain Poets and the Beat Poets shunned social convention through their experimental art forms, the Black Mountain Poets were affiliated with the Black Mountain College in North Carolina, whereas the Beat Poets were concentrated in California. No error

5. Situated at the site where the Pacific Ocean meets the San Francisco Bay, the Golden Gate spanned by its now-famous bridge , which was completed in 1937. No error

6. Although epiphytes are plants which used other plants for support, they are not parasitic because they have broad leaves that catch water as it drips through the canopy of the tropical forest. No error

7. Theoretically _ bacterial meningitis is far more serious than the viral form of the disease, even though viral meningitis, like any viral disease, can result in an infection for which antibiotics will be ineffective. No error

8. He was desperate to improve his grades before graduation ; However, he could not, due to the constant distraction from after-school activities, see any academic progress. No error

9. It is possible to predict when lightening will occur because sparks between clouds and the ground are accompanied by light before they are seen as lightening . No error

10. Because it evaluates a horse's ability to execute defined movements, the equestrian disciplines of showjumping and dressage are set in a confined area with everything required for the events in place. No error

11. By the end of 1930 , the International Astrological Union had assigned boundaries on the celestial sphere and grouped stars into 88 constellations. No error

12. John F. Kennedy was indisputably the most important US president having been assassinated during his brief tenure in the Oval Office. No error

13. DDT, an organic compound formerly used in insecticides, was withdrawn from the market because it was highly toxic and proved to have a long-lasting negative impact with the environment. No error

14. When Alexander Graham Bell invented the acoustic telegraph, little did he know that he would not only be known for his experiments with sound, but also going down in history as the father of the modern telephone. No error

15. Harmonic accompaniment, a sound sequence that make a recognizable pattern, has been understood by musicians as nearly always being subordinate to melody. No error

16. Normally grown in warm climates, many types of melons are cultivated in greenhouses nowadays, even out of their usual growing seasons. No error

17. The Hawaiian king Kamehameha _ who was descended from Kamehameha IV, abandoned the Hawaiian constitution and imparted fewer rights to his subjects, as all royal rulers did. No error

18. With books in print in more than 100 countries, Mark Twain's work has had his share of admirers', as well as an abundance of critics. No error

19. The longest river in the world, the Nile flows from its headstream to the Mediterranean delta in Northeast Egypt. No error

Instructions for Questions 20 to 30 – Select the best substitute for the **highlighted** parts of the following ten sentences. The first answer [choice A] is identical to the original sentence. If you think the original sentence is best, then choose A as your answer.

20. The defining characteristics of colleges formed during the Antebellum Period **included its** solicitation of government support, promotion of educational opportunities, and provision for educational and technical students.
    A. included its
    B. included their
    C. includes its
    D. includes their
    E. including the

21. Franz Kafka was one of a handful of European novelists **overcoming** intense self-doubt and see his work published during his lifetime.
    A. overcoming
    B. having overcome
    C. who had overcome
    D. to have overcome
    E. to overcome

22. Non-tariff barriers threaten free trade and limit a country's commerce by impeding **its ability to export, to invest, and their financial growth.**
    A. its ability to export, to invest, and their financial growth.
    B. their ability to export, to invest, and financial growth.
    C. it's ability to export, to invest, and their financial growth.
    D. its ability to export, to invest, and financial growth.
    E. its ability to export, to invest, and to grow financially.

23. The first woman elected to the US Congress, suffragist and pacifist Jeannette Rankin **was stanchly promoting** women's rights during her career.
    A. was stanchly promoting
    B. was promoting stanchly
    C. staunchly promoted
    D. having staunchly promoted
    E. having to staunchly promote

24. The diplomat **mailed the letter to the embassy containing confidential information.**
    A. mailed the letter to the embassy containing confidential information.
    B. mailed the letter to the embassy, containing confidential information.
    C. mailed, containing confidential information, the letter to the embassy.
    D. mailed the letter containing confidential information to the embassy.
    E. containing confidential information, mailed the letter to the embassy.

25. Providing an alternative to other energy sources, nuclear power addresses concerns about **increasing air pollution, decreasing fossil fuels, and helps to control costs associated with rising inflation.**
    A. increasing air pollution, decreasing fossil fuels, and helps to control costs associated with rising inflation.
    B. increasing air pollution, decreasing fossil fuels, and help to control costs associated with rising inflation.
    C. increasing air pollution, decreasing fossil fuels, and controlling costs associated with rising inflation.
    D. increasing air pollution, decreasing fossil fuels, and associated costs with rising inflation.
    E. increasing air pollution, decreasing fossil fuels, and associating costs with rising inflation.

26. **Drafting the *Declaration of Independence* and serving as Secretary of State,** Thomas Jefferson has a prominent place in US history.

A. Drafting the *Declaration of Independence* and serving as Secretary of State,
B. Drafting the *Declaration of Independence*, and he served as Secretary of State,
C. By drafting the *Declaration of Independence*, and he served as Secretary of State,
D. Drafting the *Declaration of Independence* and to serve as Secretary of State,
E. With drafting the *Declaration of Independence*, he served as Secretary of State and,

27. **Honeysuckle is a well-known species of climbing plant native to the northern hemisphere and consequently** often grown for ornamental purposes in patios and gardens.

A. Honeysuckle is a well-known species of climbing plant native to the northern hemisphere and consequently
B. Honeysuckle is a well-known species of climbing plant native to the northern hemisphere and thus
C. Honeysuckle is a well-known species of climbing plant native to the northern hemisphere and furthermore
D. Since honeysuckle is a well-known species of climbing plant native to the northern hemisphere, it is
E. Honeysuckle is a well-known species of climbing plant native to the northern hemisphere and because it is

28. **Overworked and underpaid many employees seek** help from their union representatives.

A. Overworked and underpaid many employees seek
B. Overworked and underpaid, many employees seek
C. Overworked and underpaid many employees seek,
D. Overworked, and underpaid many employees seek
E. Overworked and underpaid many employees, seek

29. With its sub-zero temperatures and frozen landscape, **hardly no one considers Siberia to be the ideal tourist destination.**

A. hardly no one considers Siberia to be the ideal tourist destination.
B. no one hardly considers Siberia to be the ideal tourist destination.
C. hardly no one considers Siberia as the ideal tourist destination.
D. no one considers Siberia hardly to be the ideal tourist destination.
E. Siberia can hardly be considered to be the ideal tourist destination.

30. **When air rises and condenses into precipitation, this phenomenon** is known as a low-pressure system.

A. When air rises and condenses into precipitation, this
B. When air rises and condenses into precipitation, this phenomenon
C. The phenomenon of air rising and condensing into precipitation, this
D. The phenomenon of air rising and condensing into precipitation
E. When air rises and condenses into precipitation, a phenomenon which

Instructions for Questions 31 to 36 – Read the draft essay below and then choose the best answers to the questions that follow.

(1) The major significant characteristic of any population is its age-sex structure, this is defined as the proportion of people of each gender in each different age group. ***(2)*** The age-sex structures of various populations have social policy implications. ***(3)*** For instance, a population with a high proportion of elderly citizens needs to consider its governmentally-funded pension schemes and health care systems. ***(4)*** A demographic with a greater percentage of young children should ensure that its educational funding and child welfare policies are implemented efficaciously. ***(5)*** The composition of a population changes over time. ***(6)*** The government may need to re-evaluate its funding priorities.

(7) Low birth rates can affect the size of the population, especially if there is a governmental policy that attempts to control demographics by restricting the number of children families can have. ***(8)*** Demographic change can also occur due to unnaturally high death rates. **(9)** After a disease epidemic or natural disaster, uncharacteristically high numbers of death become especially evident. ***(10)*** Finally, migration is another factor in demographic attrition, because in any population, a certain amount of people, may decide to emigrate, or move to a different country.

31. What is the best to do with sentence 1? Sentence 1 is provided again here for ease of reference.

    The major significant characteristic of any population is its age-sex structure, this is defined as the proportion of people of each gender in each different age group.

    A. Replace the word "this" with the word "which"
    B. Delete the word "major"
    C. Replace the word "is" with "should be"
    D. Delete the words "defined as"
    E. Delete the word "each"

32. *Which one of the following words or phrases would be best inserted at the beginning of sentence 4?* Sentence 4 is provided again here for ease of reference.

    A demographic with a greater percentage of young children should ensure that its educational funding and child welfare policies are implemented efficaciously.

    A. Just as
    B. Accordingly,
    C. In contrast to
    D. In spite of,
    E. On the other hand,

33. Which of the following sentences would be best inserted between sentences 4 and 5?
    A. Young children can be precocious or even gifted.
    B. Nevertheless, young children usually do not require significant resources.
    C. Many regard the government as capricious in its budgeting and spending decisions.
    D. In addition, a population with an increasing proportion of adults of child-bearing age may need to evaluate funding for child-birth and parenting classes.
    E. Yet, we must bear in mind that it is not feasible to provide full funding to all sectors of the population.

34. What is the best way to revise and combine sentences 5 and 6?

A. The composition of a population changes over time because the government may need to re-evaluate its funding priorities.
B. Since the composition of a population changes over time, the government may need to re-evaluate its funding priorities.
C. The composition of a population changes over time, surely, the government may need to re-evaluate its funding priorities.
D. The composition of a population changes over time so that the government may need to re-evaluate its funding priorities.
E. The composition of a population changes over time, even though the government may need to re-evaluate its funding priorities.

35. What word or phrase is best inserted at the beginning of sentence 8? Sentence 8 is provided again here for ease of reference.

Demographic change can also occur due to unnaturally high death rates.

A. Moreover,
B. With regard to
C. Precisely,
D. For the most part,
E. Although

36. What is the best revision to the underlined portions of sentence 9?

After a disease epidemic or natural disaster, uncharacteristically high numbers of death become especially evident.

A. uncharacteristically high amount of death
B. uncharacteristically high amount of deaths
C. uncharacteristically high number of deaths
D. uncharacteristically high numbers of deaths
E. uncharacteristically high figures of deaths

Instructions for Questions 37 to 40 – Read the basic research skills questions that follow and then choose the best answers.

37. Which sentence below best describes the term “primary source”?

A. a document which was written or physical evidence that was created during the time under study
B. a document which was written about the item under study
C. an article that analyzes previous research conclusions or findings
D. an article that explicates existing research conclusions or findings
E. an article that criticizes previous research conclusions

38. What is the main purpose of reviewing the list of sources in an article when conducting research for a term paper or other research project?
A. to check for plagiarism in the work of others
B. to assess the veracity of quotations within the article
C. to identify further sources to use in your research
D. to determine whether the list of references has been provided in the correct format
E. to avoid having to do extra reading on your subject

39. Smith, C. (2013). *American history in the twenty-first century*. New York: Independent Publishers.

The item above is from a list of references. Which of the following is being cited?

A. a newspaper article
B. a magazine article
C. an article from a scholarly journal
D. an online publication
E. a book

40. A student is writing a research paper on the impact of a recent change in state law on educators in high schools in his or her state. Which of the following would be the most relevant secondary source for the student’s paper?
A. A textbook on the history of state law affecting educators in high schools
B. A newspaper article in which a local elementary school teacher complains about recent changes to the law
C. An article in a periodical in which high school teachers in the state express their opinions on the law change
D. Discussions with school board members and parents who reside in the area of the local high school
E. A magazine article on recent changes in education in the United States

Praxis Core Writing Practice Test 2 – Answers and Explanations

1. You should select: the comma. Remember that a comma should not be used before "that" when forming relative clauses.

2. You should select: themselves. The subject of this clause is "any substance," so the singular form of the pronoun (itself) should be used.

3. You should select: moorish. Moorish is a proper adjective, so it needs to be capitalized.

4. You should select: no error. The sentence contains no errors in grammar or usage.

5. You should select: spanned. The bridge spans the bay, so the passive verb form (is spanned) is needed in this sentence.

6. You should select: used. The sentence is describing a scientific principle, so the present simple tense (use) is required.

7. You should select: no comma. A comma should be placed after the adverb "theoretically" since this adverb is placed at the beginning of the sentence.

8. You should select: the semicolon. The word "however" forms a new sentence because it is capitalized, so a period needs to be used here, instead of the semicolon.

9. You should select: no error. The sentence contains no errors in grammar or usage.

10. You should select: it evaluates. The subject of the sentence is "disciplines," which is plural, so the clause should use the plural form "they evaluate."

11. You should select: no error. The sentence contains no errors in grammar or usage.

12. You should select: having been assassinated. The perfect passive infinitive verb form "to have been assassinated" is needed because the word "indisputably" earlier in the sentence shows that this topic is still of current interest.

13. You should select: with. The word "impact" takes the preposition "on."

14. You should select: going. The simple present form of the verb "go" needs to be used in this sentence in order to form a parallel structure with the simple present form of the verb "be" in the "not only" clause of the sentence.

15. You should select: that make. The subject of the clause is "sequence", which is singular, so the verb "makes" is needed.

16. You should select: no error. The sentence contains no errors in grammar or usage.

17. You should select: no comma. A comma is needed before the word "who" because this part of the sentence forms a non-defining relative clause. In other words, the name of the king is provided, so this part of the sentence just gives additional information.

18. You should select: his. The subject of the clause is "work," which is singular, so the pronoun "its" should be used.

19. You should select: Northeast. The adjective should not be capitalized since it is not a proper adjective.

20. The correct answer is B. The subject of the sentence is "colleges," so the plural form of the pronoun ("their") needs to be used. The Antebellum Period is in the past, so the past simple tense ("included") is the correct verb form.

21. The correct answer is E. The base form of the verb "see" is used in the second part of the sentence, so the verb "to overcome" is needed in order to create the correct parallel structure between the verb forms.

22. The correct answer is E. The subject of the sentence is singular ("a country's commerce"). Accordingly, the singular pronoun ("its") is correct. In addition, the infinitive verb form ("to grow") is required since the preceding verb forms are in the infinitive form ("to export [. . .] to invest").

23. The correct answer is C. The sentence describes an action that was completed in the past, so the past simple form ("promoted") is correct.

24. The correct answer is D. The sentence as written has a misplaced modifying phrase and thereby suggests that the embassy contains the confidential information. The letter, rather than the embassy, contains the confidential information, so the modifying phrase ("containing confidential information") needs to be placed directly after the noun to which it relates ("the letter").

25. The correct answer is C. All of the verb forms need to be in the –ing form in order to create the correct parallel structure.

26. The correct answer is A. The sentence is correct as written since both verbs in the modifying phrase are in the –ing form ("drafting [. . .] and serving").

27. The correct answer is D. The sentence as written suggests that the plant is popular simply because it is indigenous to the northern hemisphere. By beginning the sentence with "since," the correct emphasis is placed on the climbing aspect of honeysuckle, which is what makes the plant popular in patios and gardens.

28. The correct answer is B. The introductory phrase "overworked and underpaid" describes the grammatical subject "many employees." Therefore, a comma needs to be used after the word "underpaid."

29. The correct answer is E. The phrase "With its sub-zero temperatures and frozen landscape" describes Siberia, so the word "Siberia" needs to be placed after the comma.

30. The correct answer is D. As written, the sentence contains a fragment ("When air rises and condenses into precipitation"). Answer D is the only response that corrects the fragment without improperly using the demonstrative pronoun "this."

31. The correct answer is A. While the original sentence is a bit verbose, the relative pronoun "which" is needed in order to correct the comma splice.

32. The correct answer is E. The idea of the elderly in sentence 3 is being contrasted with the idea of young children in sentence 4, so the introductory phrase needs to show a contrast. The linking phrase also needs to be followed by a comma.

33. The correct answer is D. At this point, the essay is providing examples of how the composition of a population changes over time. Sentence D is the only answer that provides such an example.

34. The correct answer is B. Answer B is the only choice that is correct grammatically and that provides the correct cause-and-effect relationship between the two ideas.

35. The correct answer is A. Sentence 8 is giving an additional example of the way in which the composition of a population can change. “Moreover” is the only answer choice that shows that an additional example is being given, so it is the correct answer.

36. The correct answer is D. More than one person would have died, so the words “number” and “death” need to be plural in this instance. A plural form is also needed because of the verb form “become”.

37. The correct answer is A. Remember that a primary source provides statistical or documentary evidence.

38. The correct answer is C. The list of sources will show other books and articles on the subject that you are researching, so you can use it to identify further sources that you can read on your subject.

39. The correct answer is E. We know that a book is being cited because the title is given in italics. In addition, a year is given, rather than a specific date, which identifies the source as a book.

40. The correct answer is C. Answer A provides a source that is much too general. Answers B and D would be biased towards the local level, rather than focusing on the entire state. Answer E provides material for the entire Unites States, rather than on the particular state. Answer C is the only response that focuses on the state-wide level.

Argumentative Essay Topic – Practice Test 2

You may wish to refer back to the section entitled "Praxis Core Essay Tips" before writing the essays in this part of the practice test.

Instructions: You are allowed 30 minutes to complete each essay in this part of the test. When you have finished, study the model essays and comments that follow each sample response.

Look at the essay topic below and then write your essay.

"Watching television programs is of no value to children whatsoever."

To what extent do you agree or disagree with the above statement? Provide reasons and examples to support your response.

Sample Argumentative Essay – Practice Test 2

Televisual media has become a pervasive force in the lives of families around the world today. Yet, a central question remains regarding whether watching television is harmful or beneficial for children. In my opinion, television programs present three major concerns in the case of children, including depictions of violence, the use of profane language, and the representation of poor moral role models.

Television programs that portray violence are a paramount concern for parents nowadays. Recent research has shown that children may commit acts of violence because they wish to emulate the behavior that they see on television. This is especially true when violent acts are committed by well-known action heroes. In addition, television programs show cartoon figures, as well as actors, committing violent acts. Using comic situations to depict violent themes causes further problems with the way in which young people view violence.

Television programs that contain profane or disrespectful language also worry parents with young children. Because censorship laws have relaxed over the past few decades, it has become very common for television programs of each and every kind to show characters expressing impolite, rude, and insulting utterances to one another. Research has shown that children unfortunately sometimes try to imitate the actions they watch on their television screens.

Finally, some parents are upset about the moral behavior depicted on television. As they struggle to teach their children moral and ethical values, parents might despair about the lack of morals and ethics represented in some of the so-called role models that their children see on the screen. For instance, certain characters not only have no remorse for their immoral actions, but also frequently go unpunished by larger society.

Because of these factors, many parents believe that television programs send their youth the wrong kinds of messages. The emulation of this poor behavior by their children is something they wish to avoid at all costs, and they have accordingly decided to ban television in their households for these reasons.

Comments on the Sample Argumentative Essay for Practice Test 2

Televisual media h*as become* a *pervasive* force in the lives of families around the world today. **Yet,** a central question remains regarding whether watching television is harmful or *beneficial* for children. In my opinion, television programs present three *major* concerns in the case of children, including depictions of violence, the use of *profane* language, and the representation of poor moral role models.

Television programs that portray violence are a *paramount* concern for parents **nowadays**. **Recent research has shown** that children may commit acts of violence because they wish to *emulate* the behavior that they *see* on television. **This is especially true when** violent acts *are committed* by well-known action heroes. **In addition,** television programs *show* cartoon figures, **as well as** actors, committing violent acts. Using comic situations to depict violent themes *causes* **further** problems with the way in which young people view violence.

Television programs that *contain* profane or *disrespectful* language **also** worry parents with young children. **Because** censorship laws have relaxed over the past few decades, it has become very common for television programs of each and every kind to show characters expressing impolite, rude, and insulting *utterances* to one another. **Research has shown that** children unfortunately sometimes try to imitate the actions they watch on their television screens.

Finally, some parents are upset about the moral behavior depicted on television. **As** they struggle to teach their children moral and *ethical* values, parents might *despair* about the lack of morals and ethics represented in some of the so-called role models that their children see on the screen. **For instance**, certain characters **not only** have no *remorse* for their immoral actions, but also frequently go unpunished by larger society.

Because of these factors, many parents *believe* that television programs send their youth the wrong kinds of messages. The *emulation* of this poor behavior by their children *is* something they wish to avoid at all costs, and they have **accordingly** decided to ban television in their households **for these reasons**.

Comments:

1) The essay is a high-level response because it contains a clear thesis statement that directly states the writer's position. The thesis statement is underlined in the preceding copy of the sample essay.

2) It displays organization and logical development of ideas by making connections between points and examples. The logical development is further supported by the use of effective sentence variety. The phrases that the student uses to organize his or her thoughts and to vary his or her sentence patterns are highlighted in **bold**. You should study the highlighted parts of these sentences and try to use similar phrases in your essay on the exam.

3) The student displays a very good grasp of academic vocabulary, which facilitates his or her use of language. See the words in *italics* for examples of the academic vocabulary used in the essay.

4) The essay does not contain errors in grammar or usage.

Source-Based Essay Task – Practice Test 2

Directions:

When paraphrasing or providing quotations from the source below, provide a citation for each source by referring to the author's last name, the title of the source, or any other clear identifying method.

Assignment:

Both of the following passages address aspects of the historical period known as the Middle Ages. Read the two sources carefully and then write an essay in which you identify the most important points from the passages.

<u>Source 1:</u>

Pellman, Terry. (2014). The historical significance of the Middle Ages. *Journal of History and Society.* 28(3), 25-32.

It is December 406 A.D. in what is now Germany. It is a bitterly cold winter, and the Rhine River is frozen. It was on this site that 15,000 warriors crossed the ice and traveled into the Roman Empire of Gaul.

This invasion, although a seemingly minor incident at the time, later transpired to be one of the most significant episodes in the history of the western world. A new historical epoch, which later came to be known as the Middle Ages, would soon be established in this former Roman Empire.

The six subsequent centuries that followed the collapse of the Roman Empire formed what we now call the Middle Ages. Even though the period of the Middle Ages has diminished in historical significance in comparison to more recent events, the demise of the Roman Empire was certainly unprecedented in the fifth century.

Although some scholars consider the epoch of the Middle Ages to be an unenlightened period, there were, in fact, many important developments during this time. Specifically, farming communities were established throughout this historical period, meaning that most people were involved in agrarian pursuits. Recent research shows that our present-day agricultural systems owe a great debt to our ancestors from the Middle Ages.

Middle Age farmers also developed several farming implements, such as the plow and the water mill. As these agricultural communities thrived, families were better able to support themselves, because the food supply was more predictable and abundant.

Due to these improvements in living standards, a great deal of population expansion also occurred during this time. This led to the establishment of a new and vital society during the Middle Ages.

<u>Source 2:</u>

Smith, Sandra. (2013). The social, economic, and political aspects of the Middle Ages. *Academic Review.* 14(2), 7-12.

As a result of the Germanic invasion in the fifth century, the autocratic system of Roman government was overthrown. In its place today is a collection of independent democratic nations. However, this development would not have been possible without the foundations laid throughout the Middle Ages.

Indeed, The Middle Ages was a time of significant social and political change. A productive process lay beneath many seemingly every-day, banal activities during this era. New societies began to materialize as the German invaders became acquainted with the Roman inhabitants. This intermingling of nationalities and ethnic groups was an important process that should not be overlooked because this type of hybridity bears a great deal of resemblance to the ethnic diversity of certain communities in modern society.

Nevertheless, economic stratification was still present at this time. Many of the invading warriors had established themselves as affluent farmers. This wealth was in stark contrast to the lives of the lower class slaves and peasants, who were often forced to live in horrific conditions.

In addition, this period also witnessed the rise in imperialism, defined as a political system in which a king or queen has absolute power. While many kings strived to rule in accordance with the law, some rulers treated their citizens harshly, without following established legal restrictions.

Despite their appalling living conditions, the common populace began to challenge the imperial system during the Middle Ages. As the attitudes of people towards their rulers changed, the balance of power in the political system began to shift. To a significant extent, these challenges influenced the functioning of present day political systems.

Sample Source-Based Essay – Practice Test 2

The Middle Ages, the period that started in the fifth century and ended in the eleventh century, has often been considered a time of social and political revolution. **Pellman states that** beginning with the invasion of 15,000 German invaders into the Roman Empire called Gaul and *culminating* in the ruin of the Empire, this era would later prove to be one of the most important in history.

Pellman repudiates the view that the Middle Ages was the **epoch** of an *inferior* civilization. He asserts that were many significant discoveries during this time, especially in the area of agriculture. He comments that "as [. . .] agricultural communities thrived, families were better able to support themselves, because the food supply was more predictable and abundant" ("The historical significance of the Middle Ages"). **Thus**, as sources of nutrition became secure, the population experienced an *unprecedented* boom.

While Pellman seems to emphasize the historical importance of the Middle Ages, **Smith focuses on** the changes to the political and economic systems during this time period. She explains that the system of government encountered challenges during Roman rule**. She also comments on** the social impact of these changes, explaining that German warriors began to integrate socially with the Romans. **She claims that** "this intermingling of nationalities and ethnic groups was an important process" because the resulting community was more diverse than it was previously ("The social, economic, and political aspects of the Middle Ages").

Importantly, Smith clarifies that there was still a vast *dichotomy* between the rich and the poor. **She highlights the fact that** some citizens, namely peasants and slaves, had to endure harsh living conditions. **In addition, she points out that** "some rulers treated their citizens harshly," bending the laws to fulfill their own purposes ("The social, economic, and political aspects of the Middle Ages").

Finally, many people decided not to accept the unfairness that their rulers imposed on them. Because of this, certain rulers lost a degree of their power. **Arguably**, this change in the power structure bears a good deal of resemblance to modern political systems.

Comments on Sample Source-Based Essay 2

1) It contains all of the main points expressed in the reading passages.

2) It is well organized. Notice the words and phrases in **bold**, which the student uses to make the summary flow well. The sources are integrated smoothly, and the student provides insights where needed.

3) The student has provided citations for each use of the references, as required.

4) The essay is grammatically correct. Notice that in summaries for this writing task, the verb in the first sentence of the essay is often in the *present perfect* tense. If the summary is asking you to describe a historical event, the verbs in most of the other sentences should be in *past simple, active or passive*.

5) The essay contains high-level academic vocabulary. Please notice the words in *italics*.

BONUS MATERIAL – ESSAY CORRECTION EXERCISES

Instructions: Read the early drafts of the essays below and choose the best answer to the questions that follow each one. The answers are given after the final question.

Essay 1

(1) Earthquakes occur on the surface of the earth when there is motion in the tectonic plates in the earth's crust. **(2)** The crust of the earth contains twelve such tectonic plates which are from four to ten kilometers in length when located below the sea. **(3)** However, those on land can be from thirty to seventy kilometers long.

(4) Fault lines, the places where these plates meet, build up a great deal of pressure. **(5)** Therefore, he two plates will eventually shift or separate since the pressure on them is constantly increasing, and this build-up of energy needs to be released. **(6)** Therefore, tectonic plates cause earthquakes, both at land and at sea.

(7) When the plates shift or separate, there has been an occurrence of an earthquake, also known as a seismic event. **(8)** The point where the earthquake is at its strongest is called the epicenter. **(9)** In addition, waves of motion travel out from the epicenter, where they cause widespread destruction to an entire geographical area.

(10) With the likelihood for earthquakes to occur, it is essential that earthquake prediction systems are in place. **(11)** The purpose of earthquake prediction is to give advanced warning to the population, thereby saving lives in the process. **(12)** Yet, these prediction systems need to be reliable in order to be of any practical use. **(13)** For this reason, scientists are currently attempting to conduct research on the probability of earthquakes along each of the twelve fault lines.

1. Which is the best version of the underlined portion of sentence 2, reproduced below?
   The crust of the earth contains twelve such tectonic <u>plates which</u> are from four to ten kilometers in length when located below the sea.
   A. Leave it as it is now.
   B. plates, which
   C. plates – which
   D. plates, that

Tip: The word "which" forms the beginning of a non-restrictive relative modifying clause. If you are unsure about this concept, please refer back to the grammar guide to learn how to use and punctuate these types of clauses.

2. Sentence 4 is reproduced below.
   Fault lines, the places where these plates meet, build up a great deal of pressure.
   The student is considering adding the following at the end of sentence 4:
   because the plates are constantly pressing on each other
   Should the student make this addition?
   A. No, because adding these words would interrupt the flow of the essay.
   B. No, because the paragraph focuses on the build-up of pressure, rather than on the plates.
   C. Yes, because it elaborates on the cause of the increase in pressure, which is discussed in the next sentence.
   D. Yes, because this event fits in with the historical account mentioned in the paragraph.

Tip: For questions on adding text to the paragraph, you need to analyze whether the additional text elaborates on the topic of the paragraph or supports claims made in the text.

3. Which is the most logical placement for sentence 6, reproduced below?
Therefore, tectonic plates cause earthquakes, both at land and at sea.
A. Where it is now.
B. After sentence 3.
C. After sentence 7.
D. After sentence 9.

Tip: This is a question on maintaining the logical sequence of information and ideas. Sentence 6 mentions earthquakes at land and at sea. To put the sentence in the best place, you need to determine which paragraph describes phenomena both at land and at sea.

4. Which is the best version of the underlined portion of sentence 7, reproduced below?
When the plates shift or separate, there <u>has been</u> an occurrence of an earthquake, also known as a seismic event.
A. Leave it as it is now.
B. was
C. are
D. is

Tip: This question assesses your knowledge of verb tense. Remember that the present tense is used when describing natural phenomena or scientific facts, as in this paragraph. You may wish to look back at the "Review of Verb Tenses" section of the grammar guide if you need to review these concepts.

5. Which is the best decision regarding the underlined portion of sentence 9, reproduced below?
In addition, waves of motion travel out from the epicenter, <u>where they cause widespread destruction to an entire geographical area</u>.
A. Leave it as it is now.
B. Change it to "causing widespread destruction to an entire geographical area."
C. Place it at the beginning of the sentence.
D. Delete it and put a period at the end of the new sentence.

Tip: This is a question on identifying problems in modifier placement. When using modifying phrases or clauses, be sure you make clear which word or phrase you are modifying. In this sentence, the waves cause the destruction. Note that the destruction does not occur in the epicenter.

Essay 2

(1) It is often said that every cloud has a silver lining. **(2)** However, when I received word that my application for college had not been accepted, I thought life as I knew it was going to end. **(3)** As I saw the upcoming academic year stretch out in front of me like a deserted highway, I mused, "How could life be so unjust?" **(4)** Little did I know that this delay in my life would ultimately lead to other events.

(5) I will never forget the moment the registrar said, "I'm really sorry, but you've missed the cutoff for this year." **(6)** Had I realized that a simple administrative error on my part was going to delay my studies, I certainly would have been more careful in submitting the necessary forms. **(7)** I was so sure that I was going to be accepted that I had not even bothered to look for work. **(8)** These are the events that led me to embark upon what could have been a year of worry and doubting myself.

(9) However, instead of sinking into a quagmire of thoughts, I decided to take that year as an opportunity to rethink my options. **(10)** I began to ask myself a hard question: "Did I really want to study for the degree program I had chosen?"

(11) I spent weeks scouring the internet for various degree programs and requested a plethora of catalogues from colleges in other states. **(12)** I then narrowed down my options to nine or ten different colleges. **(13)** I also requested financial aid and scholarship information from the colleges I had chosen. **(14)** In the end, three colleges looked the most promising, so I decided to submit applications for admissions, as well as scholarship applications to those places.

(15) The joy I felt when I found out that I had received a full scholarship more than outstripped the agony I had experienced less than a year earlier. **(16)** Had I not had that setback, I never would have decided to pursue a degree in business studies.

6. Which is the best version of the underlined portion of sentence 3, reproduced below?
As I saw the upcoming academic year stretch out in front of me like a deserted highway, I mused, "How could life be so unjust?"
   A. Leave it as it is now.
   B. unjust?
   C. unjust!"
   D. unjust"?

Tip: This is a question on end-of-sentence punctuation. Remember that a question mark should go inside the final quotation mark when the question is part of the original dialogue.

7. Which version of the underlined portion of sentence 4, reproduced below, best introduces the events described in the remainder of the essay?
Little did I know that this delay in my life would ultimately lead to other events.
   A. Leave it as it is now.
   B. life course would ultimately be such a long and winding path.
   C. academic path would ultimately lead to something truly wonderful.
   D. academic pursuits would ultimately lead to several mixed blessings.

Tip: When answering questions on the relevance of word choice to topic and focus, look for the answer that provides the best logical connection of ideas within the essay. In this question, choose the answer that best connects to the writer's academic life, which is described in the main body of the essay, as well as to the writer's receipt of a scholarship, which is described in the conclusion to the essay.

8. Which is the best decision regarding the underlined portion of sentence 8, reproduced below?
These are the events that led me to embark upon what could have been a year of worry and doubting myself.
   A. Leave it as it is now.
   B. of worry and self-doubt.
   C. of: worry and doubting myself.
   D. of worrying and self-doubt.

Tip: This is a question on parallel structure. "Parallel structure" means that you need to use the same parts of speech when you list items in a series. So, all of the items need to be nouns or verbs, for example. In this essay, we are listing two concepts, but "worry" is a noun and "doubting" is a verb.

9. In sentence 9, the writer wants to link to her comments from sentence 8. Which version of the underlined portion of sentence 9, reproduced below, best achieves that goal?
However, instead of sinking into a quagmire of thoughts, I decided to take that year as an opportunity to rethink my options.
   A. worrying
   B. depression
   C. worrying and depression
   D. self-recrimination and anxiety

Tip: When writing topic sentences for each paragraph, you need to maintain the proposition. In other words, there should be a clear flow of ideas or claims from one paragraph to the next. To find the correct answer, determine which answer choice contains synonyms for "self-doubt" and "worry" from the previous sentence.

10. Which choice best combines sentences 12 and 13, reproduced below, at the underlined portion?
I then narrowed down my options to nine or ten different colleges. I also requested financial aid and scholarship information from the colleges I had chosen.
   A. colleges. and I also requested financial aid and scholarship information from the colleges I had chosen.
   B. colleges, and requested financial aid and scholarship information from the colleges I had chosen.
   C. colleges and requested financial aid and scholarship information from them.
   D. colleges, and requested financial aid and scholarship information from them.

Tip: Be sure you know how to coordinate and subordinate sentences and punctuate them correctly for the exam. "Coordination" generally means that you are combining two or more similar ideas within a single sentence. The word "and" is often used to coordinate sentences. You also need to be sure that you know how to punctuate the sentence correctly. Remember that a comma is needed after the word before the word "and" when there is a grammatical subject after the word "and." You also need to be confident about how to subordinate sentences correctly for the exam. "Subordination" means that you are combining different ideas within a single sentence. In contrast to coordination, the ideas are not given equal weight or emphasis in a subordinated sentence; rather, one idea is generally given more emphasis. If you are unsure about these concepts, please refer back to the "Subordination" section of the grammar guide.

Essay 3

(1) In the fall of 1859, a discouraged man was sitting in his run-down law office in Springfield, Illinois. **(2)** He was fifty years old, and he had been a lawyer for twenty years, earning around three thousand dollars a year. **(3)** His tangible possessions and property consisted of 160 acres of farm land in Iowa and the house which he lived in Illinois. **(4)** Although his monetary resources were limited and he was in debt, this man would later go on to do great things for his country. **(5)** His name was Abraham Lincoln.

(6) In 1859, some of Lincoln's associates had already begun to put forward the idea that he should run for president of the United States. **(7)** He discounted the notion in his usual self-deprecating manner. **(8)** Yet, as time passed, Lincoln started to believe that his candidacy for United States President might be possible. **(9)** In order to gain the support of his political party, he began to write to influential Republican Party leaders, including Norman Judd and Richard

Oglesby. **(10)** By the end of 1860, Lincoln garnered more public support, after he delivered public lectures and political speeches in various states.

(11) The Republican National Convention took place in Chicago on May 16, 1860. **(12)** William Seward, an outstandingly popular Republican leader, was predicted to win the nomination for the Republican Party. **(13)** Seward's later purchase of Alaska was termed "Seward's folly" and made him the object of political ridicule. **(14)** Nevertheless, Lincoln won 354 of the 466 total votes for nomination. **(15)** Thus, in November, 1860, Lincoln was elected as the President of the United States.

11. Which is the best version of sentence 3, reproduced below?
His tangible possessions and property consisted of 160 acres of farm land in Iowa and the <u>house which he lived in</u> Illinois.
A. Leave it as it is now.
B. house; which he lived in
C. house that he lived in
D. house in which he lived in

Tip: This is a question on relative clauses. Relative clauses often use the words "who," "that," "which," or "where." Remember that the "in which" construction can be used as a more formal alternative to "where."

12. Which option below best combines sentences 6 and 7, reproduced below, at the underlined portion?
In 1859, some of Lincoln's associates had already begun to put forward the idea that he should run for president of the <u>United States. He discounted the notion</u> in his usual self-deprecating manner.
A. United States, a notion that he discounted
B. Unites States, he discounted the notion
C. United States – this being a notion that he discounted
D. Unites States, discounting the notion

Tip: In order to combine sentences effectively, you need to look for the repetition of ideas in two subsequent sentences. Here, the words "the notion" refer back to "the idea" in the first part of the previous sentence.

13. Which is the best version of the underlined portion of sentence 9, reproduced below?
<u>In order to gain</u> the support of his political party, he began to write to influential Republican Party leaders, including Norman Judd and Richard Oglesby.
A. Leave it as it is now.
B. Thereby, to gain
C. Whereas, to gain
D. For example, to gain

Tip: For questions on using transition words and phrases, you need to analyze the relationship between the ideas within a sentence. To carry out this type of analysis, ask yourself the following types of questions: Is the transition word or phrase introducing an example or a reason? Is the answer choice grammatically correct?

14. Which is the best version of the underlined portion of sentence 10, reproduced below?
By the end of 1860, Lincoln garnered more public support, after he <u>delivered</u> public lectures and political speeches in various states.
A. Leave it as it is now.
B. had delivered

C. was delivered
D. had been delivered

Tip: Verb voice refers to whether a sentence should be written in the active or passive voice. Active voice (e.g. – delivered, had delivered) should be used when you want to emphasize the person who is doing the action. Passive voice (was delivered, had been delivered) should be used when want to emphasize the object or outcome. If you are unsure about these concepts, please refer to the section on verb tense and voice in the grammar guide.

15. Which sentence blurs the focus of the last paragraph and should therefore be deleted from the essay?
    A. Sentence 11
    B. Sentence 12
    C. Sentence 13
    D. Sentence 14

Tip: You will be asked to delete sentences that blur the focus of the essay on your writing test. You need to determine the primary focus of the essay to answer these types of questions. Note that in this essay, the focus is on Lincoln, not Seward.

Essay 4

(1) The study of the philosophy of human nature is often regarded as an investigation into the meaning of life. **(2)** This subject deals with four key problem areas: human choice, human thought, human personality and the unity of the human being. **(3)** A consideration of these four problem areas can also include scientific and artistic viewpoints on the nature of human life.

(4) The first problem area, human choice, asks whether a human being can really make decisions that can change their future. **(5)** Conversely, it investigates to what extent the individual's future is fixed and pre-determined by cosmic forces outside the control of the human being.

(6) In the second problem area, human thought, epistemology is considered. **(7)** Epistemology means the study of knowledge, it should not be confused with ontology, the study of being or existence.

(8) The third key issue, human personality, emphasizes aspects of human life that are beyond mental processes. **(9)** It takes a look at emotional, spiritual, and communal elements. **(10)** Importantly, the communal aspect focuses on community and communication, rather than on government or the philosophy of the state.

(11) Finally, the fourth problem area, the unity of the human being, explores the first three areas more fully and asks whether there is any unifying basis for human choice, thought, and personality. **(12)** In other words, humans are inherently complex and multifaceted beings.

16. Which is the best version of the underlined portion of sentence 2, reproduced below?
    This subject deals with four key problem <u>*areas: human choice, human thought, human personality and the unity of the human being.*</u>
    A. Leave it as it is now.
    B. areas, human choice, human thought, human personality, and the unity of the human being.
    C. areas: human choice, human thought, human personality, and the unity of the human being.

D. areas; human choice, human thought, human personality and the unity of the human being.

Tip: This is a question on punctuating items in a series. Remember that the colon can be used when several items are listed in a series after it. Also remember that a comma should be used after each word in the series that is before the word "and."

17. Which is the best version of the underlined portion of sentence 4, reproduced below?
The first problem area, human choice, asks whether a human being can really make decisions that can change <u>their future</u>.
A. Leave it as it is now.
B. their futures
C. his or her future
D. his or her futures

Tip: This is a question on pronoun person and number. The word being modified is "future," which is singular. So, the possessive pronoun before the word "future" should also be singular. The pronoun refers to "a human being," so the possessive pronoun should be in the third person.

18. Which is the decision regarding the underlined portion of sentence 7, reproduced below?
Epistemology means the study of <u>knowledge, it</u> should not be confused with ontology, the study of being or existence.
A. Leave it as it is now.
B. Use a colon, instead of a comma.
C. Use a semicolon, instead of a comma.
D. Insert the word "which" after the comma.
E.

Tip: In order to avoid run-on sentences, you need to identify sentence boundaries. Here we have two complete sentences, the first beginning with "epistemology" and the second beginning with "it." Remember that a semicolon can be used between two complete sentences.

19. Which is the best version of the underlined portion of sentence 8, reproduced below?
The third key issue, human personality, <u>emphasizes</u> aspects of human life that are beyond mental processes.
A. Leave it as it is now.
B. emphasize
C. does emphasize
D. will emphasize

Tip: You will need to have agreement between the subject and verb within each sentence in an essay. Note that the grammatical subject of this sentence is "human personality," which is singular. Also be sure to use the correct verb tense.

20. Sentence 12 is reproduced below.
In other words, humans are inherently complex and multifaceted beings.
The student is considering adding the following text at the end of the sentence:
, but there must be a unity or wholeness which underlies these complications
Should the student make this addition there?
A. Yes, because it addresses a rebuttal that was mentioned previously.
B. Yes, because it links to the idea of a "unifying basis" mentioned in the previous sentence.
C. No, because it introduces superfluous information into the essay.
D. No, because it fails to explain the notions of unity and wholeness.

Tip: This question is asking you to choose the best conclusion to the essay. The topic sentence of this final paragraph focuses on the unity of the human being, so look for synonyms for this idea in the new text.

Essay 5

(1) Cancer, a group of more than 100 different types of disease, occurs when cells in the body begin to divide abnormally and continue dividing and forming more cells without control or order. **(2)** All internal organs of the body consist of cells, which normally divide to produce more cells when the body requires them. **(3)** This is a natural, orderly process that keeps human beings healthy.

(4) If cell division occurs when they are not necessary, a large growth called a tumor can form. **(5)** These tumors can usually be removed, and in many cases, they do not recur. **(6)** Unfortunately, in some cases the cancer from the original tumor spreads and effects other parts of the body. **(7)** The spread of cancer in this way is called metastasis.

(8) Research has shown that some factors greatly increase the risk of cancer. **(9)** Smoking is the largest single cause of death, from cancer in the United States. **(10)** One-third of the deaths from cancer each year are related to smoking. **(11)** Making tobacco use the most preventable cause of death in this country.

(12) Choice of food can also be linked to cancer. **(13)** Research shows that there is a link between high-fat food and certain cancers, and being seriously overweight is also a cancer risk. **(14)** Cancer risk can be reduced by cutting down on fatty food and eating generous amounts of fruit and vegetables, which would decrease the chance of getting cancer.

21. Which is the best version of the underlined portion of sentence 4, reproduced below?
If cell division occurs when they are not necessary, a large growth called a tumor can form.
    A. Leave it as it is now.
    B. their
    C. are
    D. it is

Tip: For questions on pronoun-antecedent agreement, you need to be sure that the pronoun agrees in number with the grammatical subject of the sentence or clause. Here, the grammatical subject of the sentence is "cell division," which is singular.

22. Which is the best decision regarding the underlined portion of sentence 6, reproduced below?
Unfortunately, in some cases the cancer from the original tumor spreads and effects other parts of the body.
    A. Leave it as it is now.
    B. Change "effects" to "affects"
    C. Change "effects" to "excepts"
    D. Change "effects" to "expects"

Tip: Be sure you know the differences between frequently confused words for the exam. Remember that "affect" is used as a verb meaning resulting in, while "effect" is a noun meaning outcome. If you are unsure about these concepts, please refer to the "commonly confused words" section in the grammar guide.

23. Which is the best version of the underlined portion of sentence 9, reproduced below?
Smoking is the largest single cause of death, from cancer in the United States.
   A. Leave it as it is now.
   B. death—from cancer
   C. cancer death
   D. death from cancer

Tip: This is a question on avoiding unnecessary punctuation. Remember that different rules apply to the punctuation of clauses and phrases within sentences. Please refer to the punctuation and subordination sections of the grammar guide if you are unsure about these concepts.

24. Which choice most effectively combines the underlined portions of sentences 10 and 11, reproduced below?
One-third of the deaths from cancer each year are related to smoking. Making tobacco use the most preventable cause of death in this country.
   A. smoking, which means that tobacco use
   B. smoking tobacco
   C. smoking, making tobacco use
   D. smoking, and tobacco use

Tip: This is a question on sentence fragments. A sentence fragment is not a complete, grammatical sentence because it does not contain both a grammatical subject and verb. A sentence fragment should be joined to the previous or subsequent sentence. Also choose the answer that has the best style.

25. Which is the decision regarding the underlined portion of sentence 14, reproduced below?
Cancer risk can be reduced by cutting down on fatty food and eating generous amounts of fruit and vegetables, which would decrease the chance of getting cancer.
   A. Leave it as it is now.
   B. Change "would decrease" to "decrease" to make the sentence more emphatic.
   C. Place a period after "vegetables" and delete the comma and other underlined text from the sentence.
   D. Change "would decrease" to "decreasing" to make the sentence more concise.

Tip: This question is asking you to improve the concision of the essay. To answer this type of question, you need to determine whether any ideas have been unnecessarily repeated within the sentence or essay.

ANSWERS AND EXPLANATIONS FOR THE BONUS ESSAY EXERCISES

1. The correct answer is B. The word "which" forms the beginning of a non-restrictive relative clause. This means that the clause beginning with "which" does not identify which tectonic plates we are describing, but provides extra information about the plates. Commas are needed after non-restrictive relative clauses.

2. The correct answer is C. The new clause elaborates on the cause of the increase in pressure. This concept is discussed in the next sentence of the essay: "Therefore, the two plates will eventually shift or separate since the pressure on them is constantly increasing." This new clause effectively elaborates on the topic of the paragraph and helps to maintain the flow of the essay.

3. The correct answer is B. Paragraph 1 describes the tectonic plates in the earth, as well as those at sea. Sentence 6 mentions that "tectonic plates cause earthquakes, both at land and at sea." So, to maintain the correct logical sequence, the best place for sentence 6 is after sentence 3.

4. The correct answer is D. The subject of the main clause of this sentence is "an occurrence." The word "occurrence" is singular, so the singular verb "is" must be used. We are describing a scientific fact, so the present tense is needed.

5. The correct answer is B. This is a question on the correct placement of modifiers. The action of the waves causes the destruction. The sentence as originally written suggests that the damage occurs only in the epicenter.

6. The correct answer is A. A question mark needs to go inside the final quotation mark when it is part of the original dialogue. In this essay, the dialogue consists of thoughts and questions within the speaker's mind.

7. The correct answer is C. Answer choice C is the best because it will make the essay more cohesive than the other options. The mention of "academic life" in the new clause provides a logical connection to the process of applying for college described in the main body of the essay. The mention of "something truly wonderful" connects to the outcome of receiving a scholarship, which the speaker mentions in the conclusion to the essay.

8. The correct answer is B. This is a question on parallelism. "Parallelism" means that you need to use the same parts of speech when you list items in a series. In this essay, we are listing two concepts, but "worry" is a noun and "doubting" is a verb. Answer B is the best because "worry" and "self-doubt" are both nouns.

9. The correct answer is D. "Self-recrimination and anxiety" are near synonyms to "self-doubt" and "worry" from the previous sentence, so these words would give the essay the best logical structure.

10. The correct answer is C. Answer choice C is more concise than the other choices. It also does not have an unnecessary comma like the other choices. No further comma is needed because this new part of the sentence does not use the word "I" again as a grammatical subject.

11. The correct answer is D. Lincoln was living in the house, so the house is the object in the clause. Therefore, we need to use the "in which" relative clause construction. This also helps us to avoid ending the sentence with a preposition.

12. The correct answer is A. The words "the notion" need to be immediately after the comma because these words refer back to "the idea" in the first part of the sentence. So, we can use a comma to join the new clause to sentence 6.

13. The correct answer is A. "In order to" is the best transition phrase because it describes the reason why Lincoln began to write to others. We use the phrase "in order to" to show cause and effect in this way in a sentence.

14. The correct answer is B. We are placing emphasis on Lincoln as a person, so we need to use the active voice. There are two verbs or actions in the sentence: garnering support and delivering speeches. The support occurred as a result of the speeches, so the speeches came first. We normally need to use the past perfect active tense (had delivered) for the action that occurred first. The past simple active (garnered) is used for the second action.

15. The correct answer is C. The focus of the entire essay is on Lincoln, not Seward. So, the additional information about Seward is not necessary.

16. The correct answer is C. The colon placement is correct since several items are listed in a series after it. Also remember to put a comma after each word in the series that is before the word "and."

17. The correct answer is C. The grammatical subject of the clause is "a human being," which is singular. So, "his or her future" should be used in order to have agreement between the pronoun and antecedent.

18. The correct answer is C. Here we have two complete sentences, the first beginning with "epistemology" and the second beginning with "it." A semicolon can be used between two complete sentences in this way. The use of the semicolon in this way avoids inappropriate run-on sentences.

19. The correct answer is A. The grammatical subject of this sentence is "human personality," which is singular. The present tense is needed since we are discussing facts and theories.

20. The correct answer is B. The topic sentence of this final paragraph focuses on the unity of the human being, so the mention of "unity or wholeness" in the new text provides support for the main idea of the paragraph.

21. The correct answer is D. The subject of the sentence is "cell division," which is singular. So, we need to use "it is" instead of "they are."

22. The correct answer is B. In this sentence, "affect" is used as a verb meaning resulting in, while "effect" is a noun meaning outcome.

23. The correct answer is D. The words "from cancer in the United States" form a phrase, not a clause. Therefore, no comma is needed.

24. The correct answer is A. You may be tempted to select answer choice C, but answer choice A is better for stylistic reasons as it correctly joins the clause to the sentence, while avoiding two words ending in "ing" together being placed together.

25. The correct answer is C. For better concision, the words "which would decrease the chance of getting cancer" should be deleted as they repeat the notion that "cancer risk can be reduced," which is mentioned at the start of the same sentence.

Made in the USA
Coppell, TX
05 November 2026

11315623R00109